Hidden Factors
in Technological Change

Hidden Factors in Technological Change

Editors:

EDWARD SEMPER (*joint Executive Editor*)
PHILIP COGGIN (*joint Executive Editor*)
HARRY ARMYTAGE
MICHAEL BASSEY
JACK MEADOWS

Published on behalf of the Standing Conference on Schools' Science and Technology by

PERGAMON PRESS

OXFORD · NEW YORK · TORONTO · SYDNEY
PARIS · FRANKFURT

U. K.	Pergamon Press Ltd., Headington Hill Hall, Oxford OX3 0BW, England
U. S. A.	Pergamon Press Inc., Maxwell House, Fairview Park, Elmsford, New York 10523, U.S.A.
CANADA	Pergamon of Canada, Ltd., P.O. Box 9600, Don Mills, M3C 2T9, Ontario, Canada
AUSTRALIA	Pergamon Press (Aust.) Pty. Ltd., 19a Boundary Street, Rushcutters Bay, N.S.W. 2011, Australia
FRANCE	Pergamon Press SARL, 24 rue des Ecoles, 75240 Paris, Cedex 05, France
WEST GERMANY	Pergamon Press GMbH, 6242, Kronberg-Taunus, Pferdstrasse 1, Frankfurt-am-main, West Germany

First edition 1976

Library of Congress Catalog Card No. 76-13333

Printed in Great Britain by Express Litho Service (Oxford).

Contents

Acknowledgements

A discussion paper prepared by the School Technology Forum Sub-committee on General Studies and published in 1974 led to the idea of producing a book of articles on the educational implications of living in an advanced technological society. Invitations were extended to possible contributors: the response was immediate and positive, and the number grew with almost embarrassing vigour. Speed of publication was considered by all to be the overriding factor: there was no time for a nation-wide review of all possibilities; all offers of help were gladly accepted. That this organic growth of interest should have brought together teachers from the north and the south as well as from comprehensive, grammar, and independent schools and that these in turn were joined by people from Higher Education and Industry convinces me of both the need and the possibility of a common culture.

As chairman of the various discussion groups involved, I should like to express my deep gratitude to: the many who made verbal and written contributions in the early stages and in particular Mr. W. R. A. Ellis of the Geographical Association, Mr. R. H. Gammon, Director of the Southern Science and Technology Forum, Dr. Frank McKim of Marlborough College, Mr. T. C. McNeill, Headmaster of Portsmouth Technical High School, and Dr. D. T. Witcombe, Headmaster of Nottingham High School: to the fifty-eight contributors, most of whom were recommended by various members of the STF Sub-committee, and all of whom so readily accepted the suggested limit of 1000 words which, although loosely interpreted, nevertheless imposed a considerable constraint upon authors wishing to refer to case-histories or developing insights on aspects of their topics: all those who helped in any way with the preparation of the book; in particular, Professor Harry Armytage, our indefatigable consultant editor; Professor Jack Meadows of Leicester University who so willingly took on the responsibility of editing the first part of the book; Dr. Michael Bassey who kindly edited the section on simulation in schools; Mr. Edward Semper who shouldered the lion's share of the work of negotiation and assembly and to whose tireless efforts over many years in the field of school technology this book is both a testimony and a tribute; Trent Polytechnic; PETTs Interdepartmental Committee (Project—Engineers and Technologists for Tomorrow), a Government/Industry/Profession group for schools, for secretarial support; and finally to those, named and unnamed, who have so willingly associated themselves with the cause of promoting technological sensibility and literacy for all.

PHILIP A. COGGIN
Chairman, School Technology Forum
Sub-committee on General Studies

Introduction

PHILIP COGGIN AND EDWARD SEMPER

The articles in this book are written by people who range from a Vice-Chancellor of a city university to a drama adviser in rural Wiltshire, from a bishop to a humanist, and from astronomers, botanists and chemists to electronic engineers, inventors, industrialists and industrial research scientists, psychologists, town-planners, wives and, not least, women teachers. All of these have spared time from very busy occupations to affirm the need for a positive stance towards the technological society in which we live and to suggest ways in which teaching programmes and learning situations in schools might be appropriately reoriented.

Like the famous paintings of Hokusai, 'The Thirty-six Views of Mount Fuji', these articles reflect the many different viewpoints from which contributors approached their task. We make no apology for using a Japanese analogy, for Japan provides both a warning and an example. The spectacle of widespread pollution and materialism has led one Japanese scientist to call for an industrial culture in Japan that would be "warm and humane, to replace the cold inhuman western import" (*Guardian*, 24 June 1971). On the other hand, Hermann Kahn, one of the shrewdest secular prophets of our time, sees Japan as 'the most forward looking and future-oriented people in the world', fast acquiring a 'giant economic, technological and financial stature' and very likely to become both 'financially and politically powerful in international affairs'.

Meanwhile, what is happening to the Western World? In spite of contriving to put a man on the Moon, replacing human limbs and organs with operational replicas, sustaining faltering hearts with nuclear powered pace-makers, and computerising and automating shunting yards and the production of metals, there is some evidence of a loss of nerve, and of a rejection of science and technology. Many able and ambitious boys and girls want to become solicitors, accountants, doctors, civil servants, local government officers, teachers, TV personalities, research workers, but few are attracted to careers in manufacturing industries. People want to enjoy the fruits of technology—colour TV sets, deep freezers, central heating systems, telephones, motor cars and hand calculators—but seldom do they wish to make them.

Why does production excite us less than our industrial rivals? Why are so few young people in our country interested in applying knowledge, ingenuity, inventiveness and artistry? Is the retreat from science and rationality an indication that we are content, as Dr. J. Bronowski put it, that 'the next step in the ascent of man shall be taken by people elsewhere?' Bronowski hoped not but he had little doubt that 'we are being weighed in the balance at the moment'.

Some articles in the book examine current trends in technological innovation and design, and the likely social, aesthetic, environmental, economic, moral and political responses. Other articles explore ways in which schools might begin to develop in their pupils critical, informed and constructive attitudes to the many opportunities and problems associated with technological change, and thereby help to make technology more accountable to society and more responsive to the wishes of ordinary people.

Hopes of this kind have been raised before: the Second Charter of the Royal Society in 1663 called on Fellows to apply their studies to the glory of God the Creator and the advantage of the human race. Unhappily, the narrower, divisive element in puritanism rather than the creative spirit tended to dominate the subsequent history of the industrial revolution and to fragment society. The stream of inventions and industrial developments of the eighteenth and nineteenth centuries were largely the achievements of self-taught men. In consequence, like the ancient Chinese scholars who grew their fingernails long to show that they had not been contaminated by manual work, academics tended to regard technology at worst as one of the inhumanities and at best as an impure science. Moreover, the business world became as suspicious of education as education was of industry.

This threat to cultural cohesion led people like Lord Snow and Lord Ashby to question the objectives of an education which relies heavily on the disinterested pursuit of knowledge and the transmission of traditional insights, ideas, and values. Others, carried along by the Consumer Society, see education either as an investment which should yield a return by way of the skills, techniques, and knowledge which society requires, or as a consumer good, providing a service that responds to the demands of its customers, the students. These objectives need not be mutually exclusive. They can and must be reconciled. The quality of human life can only evolve out of the interactions between the objectives we set ourselves and the practical means which industry provides for their achievement.

The sands are running out. Culturally, industrially and educationally speaking 'the writing has appeared on the wall but it is no man's professional business to read it and take warning', as Dr. D. J. Price warned a quarter of a century ago. The issues at stake affect the lives of everyone. Nothing less than the active participation of all sections of society, national leaders as well as the man in the street, will be enough to ensure that the writing is accurately interpreted and effectively turned into action.

The aim of this symposium is to initiate the debate, to identify the problem, and point to possible ways ahead, rather than to propose a neat and comprehensive solution. Today's questions will be different from those of tomorrow but today's decisions will critically influence the lives of our children and our children's children.

PART I

Viewpoints on Some Technological Innovations

Introduction

Any major development in technology inevitably influences the lives of individuals and the workings of societies in both predictable and unpredictable ways. In this section, we shall be examining the wider implications of technological innovation from a variety of different viewpoints. We can lead into this discussion by pointing out the wide range of questions that might be asked about the influence of just one twentieth-century technological advance—television. What effect, if any, has the presence of television sets in innumerable households had in changing social aspirations? Has the immediacy of contact with world events, provided by television, affected our ideas concerning political activity? Does the success of television programmes on architecture and the visual arts imply any general raising of the level of aesthetic sensibility? Questions such as these can be multiplied without difficulty, and, although hardly open to simple answers, can still be discussed fruitfully.*

As will become evident in the next few pages, a thread of doubt runs through much contemporary discussion of technology. This derives, in basis, from a feeling that technological growth may often work to the disadvantage of the community as a whole. An example, frequently given, is the increased pollution that has resulted from technological expansion. Engineers and their supporters tend to feel suspicious of much of this questioning: they ask, with reason, how many people would wish to dispense with the advances that technology has introduced. But a concern for the broader implications of technological innovation need not imply an antagonism to technological activity as such. Indeed, it would be a disaster for the future of technology if the only people to examine the broader aspects of technology were those who were already opposed to its further development. It is vital that an intelligent,

* See, for example: Raymond Williams, *Television: Technology and Cultural Form* (Fontana/Collins, 1974).

sympathetic, but informed, questioning of technological activity should become widespread. In particular, future engineers should be trained to consider the wider significance of introducing new technology; for they will certainly need to justify it before the world at large.

We cannot understand the process of change in which we are involved if we limit ourselves to thinking of the democratic, industrial, and cultural revolutions as separate processes. Our whole way of life, from the shape of our communities to the organisation and content of education, and from the structure of the family to the status of art and entertainment, is being profoundly affected by the progress and interaction of democracy and industry, and by the extension of communications.

Raymond Williams

From *The Long Revolution*, Penguin, 1965 (73), pp. 11–12

Economic

CHRISTOPHER FREEMAN

Professor of Science Policy and Director of Science Policy Research Unit, University of Sussex

Of all the aspects of technology that deserve examination, the financial implications come first. On the one hand, the level of technological development in a country is strongly correlated with its economic status: on the other, the amount of funding at risk increases rapidly with the degree of technological sophistication. A.J.M.

Society supports scientific research for many different reasons. Amongst them are considerations of national prestige, military defence, education and the pursuit of knowledge for its own sake. But it is probably true to say that in the twentieth century economic efficiency and war have been the dominant objectives of national policies.

Economists have no more right than anyone else to lay down the goals of society. These depend on values held by the members of society, which are changing over time and which may conflict. It is in the political arena that these values find their expression and that these conflicts are resolved, however imperfectly. Thus one society may decide at a particular time to give a very high priority to national prestige, or to the welfare of poorer countries, or to preparation for war, or to any other of the many different objectives pursued by national governments. Economics *per se* has nothing to say about the merits of these alternatives.

But since resources are limited, and usually not all objectives can be achieved simultaneously, choices must be made and the *cost* of various alternatives is almost always relevant to decision-making about these choices. Economics has a great deal to say about costs. It differs from accountancy in not being merely concerned with money expenditures, but with real resources—labour, machinery and materials. The most fundamental concept in economics is opportunity cost—the alternatives which are foregone if resources are used in one way rather than another.

Society may choose to invest large resources in space programmes or in supersonic aircraft. But economists have the responsibility to point out that these resources are then not available to improve the standard of living in poor countries, or to improve the competitive efficiency of the mechanical engineering industry, or to improve urban transport systems, or for whatever other objectives may be competing for scarce resources.

Economics is very much concerned with reducing costs in this sense, since to produce goods or services at a lower real cost in resources means that these resources may then be made available for alternative uses, provided full employment policies are pursued. If resources were not scarce, then there would be no need for economics, which is the science of measuring and comparing the alternative costs and benefits of using resources in one way rather than another.

In practical terms, economists have been very much concerned with growth, that is to say, with ways and means of using resources more efficiently to produce a given output with fewer resource inputs. Productivity growth, whether in industry, agriculture or services, releases resources for alternative uses and permits the total product and income of society to increase. Economics has thrown a great deal of light on the ways in which industrial, government and university R & D have contributed to productivity growth.

Economists have been concerned with growth problems because most people in most societies have been very much preoccupied with improving their standard of living and still are. The very success of growth policies, through the application of science and technology to production, leads to a greater possibility of other objectives taking priority over growth, such as space exploration or the ballet and opera. For most people in the world, however, elementary improvements in living standards still tend to take the highest priority. Consequently, economists have been very much concerned with problems of underdevelopment and poverty and the international redistribution of wealth and of scientific and technical resources.

It is in this context that many economists have been sceptical of the merits of manned space programmes as a sensible use of scarce scientific and technical resources, and even more sceptical of the claims of the aircraft industry. It is true to say that no reputable economist, indeed scarcely any economist, on either side of the Atlantic, could be found who would produce any kind of economic justification for supersonic civil transport (Concorde). Not only has this Anglo-French project used more resources than any other single technical project since the war, but there is also no conceivable possibility of any economic return on the investment. The development costs (well over £1000 m) have been provided entirely by the two governments, without any prospect of recovery. But more than this, not only the production, but even the operation by national airlines, will have to be heavily subsidised by governments. This is the topsy-turvy economics of 'Alice-in-Wonderland'.

It seems in fact that no economic assessment of Concorde of any kind was made before 1964, i.e. several years after the project commenced. Once an assessment was made, it was so obviously adverse that cancellation was the clear imperative. But by this time the high priority given by De Gaulle to French national prestige and by successive British governments to not offending the French (so as to get into the EEC) ensured the continuation of this extraordinarily wasteful, although technologically satisfying, project.

Not only did Concorde defy all rational economic criteria, it also conflicted with many environmental objectives, such as noise reduction. Its persistence can be explained only in terms of the power of a particular lobby and its influence on two governments. Space research is quite a different problem, as it represents a particularly expensive, but extraordinarily interesting, type of fundamental research. Concorde, on the other hand, does not contribute to basic science. It represents only an expensive, but fast, mode of travel for a few people.

It would be quite wrong to conclude from this example that economists are the inveterate opponents of any ambitious and expensive technological project, or any long-term scientific objectives. In fact, economists have provided the most sophisticated justification for heavy investment by society in fundamental research. On the whole, economists have argued that society is under-investing rather than over-investing in fundamental research. Although the returns from basic research are very long-term and very uncertain, historically they have been extraordinarily important. The same is not true of expensive military and civil aircraft projects, despite the arguments of the industry about spin-off.

Non-economists are quite justified in pointing out that economics has often no means of measuring 'intangible' benefits such as national prestige. It is only by rating these extraordinarily highly that Concorde could be justified. Even here it is open to the critic to point out that an expensive commercial failure does little to enhance prestige, whatever its technological merits. The economic case for space research is very much stronger but economists would query how far *manned* space flight (which increased the cost enormously) was necessary to get the scientific results, and how far the *manned* space programme was motivated by other (military and prestige) factors.

When we come to consider the economics of nuclear power, the position is by no means so straightforward. Military and prestige factors have been extremely important in nuclear research and development, and are still very important factors influencing national policies. But there is a straight economic justification for the long-term development of nuclear power based on the certainty of ultimate fossil fuel depletion and the probability that other sources of energy will not be so quickly or cheaply available. Since energy supplies are fundamental for all industrial and agricultural production, this is an extremely powerful argument. However, economists would distinguish between the arguments for R & D into new sources of energy, including full-scale prototypes, and the arguments for large-scale investment. The case for a massive long-term R & D programme is very strong, but it is just as strong for solar energy as for nuclear power. It makes good economic sense to use R & D as a long-term insurance policy for society.

The case for normal commercial investment is a totally different one. As long as coal-fired power stations (or oil-fired) can produce electricity more cheaply than nuclear, and as long as energy conservation measures are in their infancy, the case for a crash construction programme in nuclear power is a weak one in economic terms, and could be disastrous in its environmental consequences. It may be justified only in terms of strategic considerations to avoid dependence on imported oil.

From all of this it is apparent that economics cannot provide any simple answers. It is an important element in most decision-making, but it must be used in combination with many other disciplines and policy considerations in arriving at decisions on scientific and technological projects. The development of good techniques of social cost-benefit analysis is a challenge to all disciplines and not merely to economics.

A technology for creating a bearable equilibrium for a very long time, commensurate with the expected biological life span of the human race, does not exist yet. We must create it. We have roughly the time scale of, say, two generations for the preparations, which must start at once, not only to avert the catastrophe, but to avert the anticipation *of a catastrophe. Our free economy relies, almost entirely, on optimistic anticipation of the future. It is this which induces people to save, and to invest in progressive industries.*

At the moment, in our present civilization, hope is synonymous with growth. We must detach the two notions before it will be possible to substitute the idea of excellence and quality of life for continuous growth . . .

Science and technology, by themselves, are of course unable to make the transition because it means, at the same time, social reforms. The reform of our economics, the reform of our values, which must go pari passu *with the development of technology . . . Pollution, avoidance of waste, and the technology of equilibrium. These are three steps of steadily increasing importance and difficulty, but they also contain increasing challenges for creative individuals.*

Professor Dennis Gabor
Nobel Prizewinner for Physics

From the Eighteenth Fawley Foundation Lecture, The Proper Priorities of Science and Technology, *Esso Magazine*, Vol. XXI. No. 4. Autumn, 1972. pp. 23–24

Social

PHILIP GUMMETT

Department of Liberal Studies in Science, The University of Manchester

Apart from its influence on the standard of living, technology has produced a fundamental change in the social structure of industrialised nations. Many of the immediate consequences of industrialisation, for example the concentration of population into conurbations have themselves led on to effects of major importance for our form of social organisation. Yet the most significant consequence of technological development may well have been its tendency to induce in us an expectation of continuing social change. A.J.M.

In all social activity, a sense of history adds to the quality of contemporary experience. This starting position cannot be fully supported here, but a simple example may help to make the point. Much concern is expressed these days about the depletion of non-renewable resources which, according to some experts, is occurring at an exponential rate so that ever larger quantities are used up in ever shorter periods of time. These experts may arrive at gloomy prognoses by extrapolating current trends into the future. Similar extrapolations could have been made in 1890 about the quantity of horse manure to be cleared from the streets of London. The growth of this nuisance could have been shown to be alarming by about 1930. However, the motor car soon began to displace the horse-drawn vehicle, and a different nuisance began to grow.

The point is not that technology will always solve our social problems; optimism about the future requires firmer foundations than that. Nor is it that extrapolations are useless, for it can be useful to consider what would happen along one dimension of social life, all other things remaining equal. Nor is it that history teaches us lessons which, in any simple sense, can be applied to contemporary problems; by definition, the subject matter of history is unique. Rather, as Lynn White, Jr. has observed in connection with the use of systems analysis in technology assessment, the point is this: 'History can offer no solutions, but it may help to guide an acute mind toward kinds of questions that in the present state of systems analysis tend to be overlooked.'(1) That is, a sense of history may help us to shape better the questions for which we must seek contemporary solutions.

When we consider the social consequences of technological innovations, our minds often turn to such areas of high technology as space programmes, supersonic transport, atomic power and some aspects of computer technology. There are important social issues associated with these, some of which are touched on elsewhere in this book, but for the moment I want to suggest that such areas of technology are remote from the experience of most people, and that for most people the *significant* effects of technology are those which they encounter personally in their daily lives. That such technologies are often mundane is irrelevant to the question of their social significance. As White, again, has observed, the invention and diffusion of the chimney in Northern Europe in the ninth to twelfth centuries led to major changes in social relations. Whereas in the days of the central fireplace, lord and lady lived, ate and slept in the same hall as their servants, so that all the strata of society mixed freely, the new flexibility of heating made possible by the chimney led to compartmentalisation. To the remark that 'the chimney may have affected the art of love more than the troubadors did', White adds that it may also have fostered the individualism of the later Middle Ages more than all the humanists. Yet, he adds, 'a high social price was paid for the new ideal of the idiosyncratic person. As communication between classes decreased, class consciousness and snobbery grew.'(2) The fact that, at the time, people may have been unaware of the social effects of this innovation only strengthens the case for the development in each individual of his sense of history.

Here, then, is the challenge. Can someone at school learn to appreciate the social effects of the changing pattern of technologies which impinge directly on the lives of ordinary people? A possible approach to this challenge is outlined below, and requires few resources beyond such standard reference works as encyclopaedias of all kinds.

Let each child compare the conditions of modern life with those which his parents experienced, say 25 years ago; which his grandparents experienced, say 50 years ago; and which his great-great-grandparents experienced, say 100 years ago. A start could be made on the 25- and 50-year periods by questioning parents and grandparents; and on the 100-year period by consulting encyclopaedias, early photographs, novels, old newspapers, or whatever else might be to hand. The emphasis in these comparisons would be upon the technology available to people at those times in their daily lives, leading to reflection upon the extent to which technological change has stimulated social change.

Typical starting-points for such inquiries might be as follows. What was the home like? What were the furniture and fittings made of? How was food stored and prepared, and where did it come from? Was there electricity, gas, water, main drainage? Were these facilities available at the time? How was the house cleaned, and the cooking utensils? How much time did people have for leisure, and how did they spend it? How did they travel, and what range of transport technologies was available? What work did grandparents do 25 years ago, and what did *their* parents do? What tools or other equipment did they use? How easy were these tools to use?

There is nothing to gain by labouring the obvious, and the approach indicated here is obvious enough once it has been suggested. If, by such inquiry, children can develop a feel for the style of life in different periods of recent history, and can relate changes in those styles—be they for good or for ill—to technological developments, then much will have been achieved. If such reflection upon the past leads to a citizenry more capable of reflection upon the present, then the achievement will be considerable. As Raymond Aron has observed:

> It may be that the technological revolution of our time has opened an unprecedented phase of the human adventure, in which the patterns supposedly observed over thousands of years—from the era of the first European empires to the detonation of the atomic bombs that devastated Hiroshima and Nagasaki—will never again be re-enacted.
>
> But the same revolution forces us to recognise the limits of our knowledge, the uncertainty of our foresight, and the possibly intrinsic uncertainty of mankind's future. Thus constrained to be modest enough to confess our ignorance, we are less likely to deceive ourselves. Our ignorance is, so to speak, well grounded, since it is acknowledgement at once of something we cannot know in the present state of human history, and of all that we have learned from the past.(3)

References

1. Lynn White, Jr., Technology assessment from the stance of a medieval historian. *American Historical Review*, **79** (Feb. 1974), pp. 1–13, at p. 3.
2. *Ibid.*, p. 9.
3. Raymond Aron, *Progress and Disillusion*, p. 295, (Pelican, 1968).

Acknowledgements

I wish to thank Drs. K. Green and G. Price for helpful conversations during the composition of this essay.

The study of great civilizations in which modern science and technology did not spontaneously develop obviously tends to raise the causal problem of how modern science did come into being at the European end of the Old World, and it does so in acute form. Indeed, the more brilliant the achievements of the ancient and medieval Asian civilizations turn out to have been the more discomforting the problem becomes. During the past thirty years historians of science in Western countries have tended to reject the sociological theories of the origin of modern science which had a considerable innings earlier in this century. The forms in which such hypotheses had then been presented were doubtless relatively crude, but that was surely no reason why they should not have been refined. Perhaps also the hypotheses themselves were felt to be too unsettling for a period during which the history of science was establishing itself as a factual academic discipline. Most historians have been prepared to see science having an influence on society, but not to admit that society influenced science, and they have liked to think of the progress of science solely in terms of the internal or autonomous filiation of ideas, theories, mental or mathematical techniques, and practical discoveries, handed on like torches from one great man to another.

Dr. Joseph Needham, F.R.S.
Master of Gonville and Caius College, Cambridge

From *The Science of Science*, edited by M. Goldsmith and A. Mackay, Pelican, 1966

Political

TOM STONIER

Professor, School of Science and Society, University of Bradford

Technological advances have been affecting the course of politics for quite some time: gunboat diplomacy in the nineteenth century, for example, would hardly have been viable had technology failed to provide gunboats. What has changed in recent years has been the scale of this effect. It is almost impossible now to find any significant aspect of diplomacy that is not influenced in some way by technology. Since World War II, the major political shadow has, of course been cast by the atomic bomb.

A.J.M.

The idea that matter was made up of small, discrete particles called 'atoms' dates back to the ancient Greeks. The nineteenth century produced new theories which provided the conceptual framework necessary for understanding chemical reactions. The nineteenth-century atom, however, did not prove to be the ultimate particle. During the first third of the twentieth century, a period of physics characterised by a great intellectual excitement, physicists began delving into the mysteries of sub-atomic structure. During this time, virtually all the work was carried out in the seclusion of universities or of special research institutes. What awareness of these developments did exist among government and industry was largely peripheral. Then, shortly before the outbreak of World War II, Otto Hahn and Fritz Strassmann, working at the Kaiser Wilhelm Institute for Chemistry in Berlin, bombarded uranium with neutrons and discovered a radioactive isotope among the products which indicated that the uranium atoms had been split. Such was the ferment in the physical sciences at that time that Philip Abelson, a young graduate student at Berkeley, California, was also pursuing a line of research which, in a few weeks would almost certainly have led him to the discovery of nuclear fission. Within about a year, scientists throughout the world had published nearly 100 articles on the phenomenon of fission. Most of the world's great centres of physics became involved.

In the meantime, committees of scientists had been formed to alert government officials to the new developments and their implications. The U.S. Navy was immediately interested, not so much in the explosive potential of uranium, but in a source of power for its submarines. Other branches of the American government varied in their response, but all ultimately became convinced and began supporting the research. At one point a committee of atomic scientists consisting of Leo Szilard, Eugene Wigner and Edward Teller was rebuffed by an Army colonel, who lectured them on the fact that it usually took two wars to develop a new weapon, and it was morale, not new arms, that brought victory. Wigner countered with the opinion that if arms were so unimportant perhaps the Army's budget ought to be cut by 30 per cent. The group got its money.

The decisions on the part of various governments to commit resources to atomic research, culminated in the American Government's decision to organise, what became known as, the 'Manhattan Project', a project ultimately consuming about 2000 million dollars. Never before in history had anyone engaged in such a massive research and development programme. Needless to say, the political decision to commit vast resources to atomic weapons research was motivated by the fear that the Germans, who were also conducting atomic research, might be ahead. Thus, the American, British, Canadian and other Allied governments had shifted completely in their attitudes towards the physicists. What was earlier considered to be merely an academic matter, had become a matter of life and death.

With the influx of massive resources, the pace of discovery and problem-solving continued to accelerate. In spite of many complex engineering problems, such as preparing sufficient quantities of uranium-235, a nuclear reactor was constructed beneath the West Stands of Stagg Field at the University of Chicago. On 2 December 1942, in this reactor, man first initiated a self-sustaining nuclear chain reaction and controlled it.

Two and a half years passed. Then 'at 0530, 16 July 1945 in a remote section of the Alamogordo Air Base, New Mexico, the first full-scale test was made of the implosion-type atomic fission bomb. For the first time in history there was a nuclear explosion. And what an explosion. . . . The test was successful beyond the most optimistic expectations of anyone. . . .' So reads a memorandum from General Leslie R. Groves to the American Secretary of War.

Just 3 weeks later, an atomic device containing uranium-235 as the fissionable material was first employed as a military weapon. The first atomic bomb was detonated at approximately 1850 feet above Hiroshima on 6 August 1945. On 9 August an atomic device containing plutonium-239 as the fissionable material was detonated at approximately 1850 feet over Nagasaki.

After the conclusion of the war, there emerged a new rivalry between the Western Block nations and the Soviet Union—the Cold War. At the same time, technological advances in the area of nuclear physics allowed scientists to proceed to the next step: an explosive nuclear fusion reaction—the hydrogen bomb—a device a thousandfold more powerful than the atom bombs used towards the end of World War II. Technology had set the stage for political decisions.

We do not know exactly what went into the decision-making process in the Soviet Union. We do know that by November 1953 the U.S.S.R. had detonated such a hydrogen bomb. We also know that within the U.S., fierce, often acrimonious, debates took place in government science circles, most notoriously those with Edward Teller on the one side and Robert Oppenheimer on the other. This debate later led to Oppenheimer's fall from governmental grace. The Teller faction won. On 1 March 1954, a 20-megaton bomb was exploded on Bikini atoll in the Pacific. Both sides had entered the thermonuclear age.

The rivalry between the two super powers and their allies, which led to the creation of super bombs, was paralleled by a corresponding arms race in delivery systems. When in 1956, the technological advances made by the Soviet Union expressed themselves in the form of the first Sputnik, the political power relationships between the Western and the Soviet Blocks changed radically: the North American heartland was no longer invulnerable to Soviet nuclear strikes. To the European allies of the U.S. the question arose whether the U.S. would still risk nuclear devastation in the event of an attack on Western Europe. The American response involved a number of political decisions, most notable among these were (1) the development of a civil defence programme which was to demonstrate to its allies that America was prepared to put its civilian population at risk, and (2) a serious effort to close, what was later referred to as, the 'missile gap'.

Both decisions required technological implementation. In addition, in order to achieve an effective civil defence programme, it was necessary to elicit the co-operation of the public at large, and begin the construction of both private and public fall-out shelters. This process initiated a public debate, with scientists entering into the debate on both sides of the issue. Within half a decade the public had become thoroughly informed about the realities of nuclear war, and political leaders were compelled to recognise that a new politics had been forced on the international system. The balance of terror achieved by the nuclear missile technology meant that both sides were stalemated.

It is in the light of the nuclear stalemate that one must look at the inability of major powers to engage in any form of effective military action with impunity. The U.S. could not risk escalating to the level of nuclear confrontation during its military involvement in South-east Asia, nor could the European Powers possessing nuclear weapons impose their will on Arab countries threatening their oil supplies. After the advent of the new military technology, the international system could never be the same.

There were other effects. The development of the sophisticated arms technology including missiles, spurred on developments ranging from electronics to polymer chemistry. Civilian applications such as the transistor and teflon brought with it an increasing specialisation and diversification of industry as well as an increasing mechanisation and automation. Society entered a new era, referred to by various authors as the 'post-industrial', 'neo-modern' or 'techtronic' society and which we will call the 'Communicative Society', because of the rapid expansion of the communications and transportation technology which integrated global social and political systems.

Characteristic of the Communicative Society is the trans-national network of economic interactions and interdependence. Technology has produced new economic patterns, which have, in turn, forced new political decisions. France and Germany, traditionally rivals, have become each other's largest trading partner. Public opinion polls in the two countries have shown a dramatic shift away from the traditional emnity and hatred towards a more sympathetic view, one even laced with friendship and admiration. The trans-national economic network of the technologically advanced countries is forcing upon them a new trans-national politics.

A few hundred years ago the Mercantile revolution produced an economy which made the feudal state obsolete and gave rise to the nation state. Today we observe a similar pattern happening in Western Europe and other areas of the world in which the electronic revolution has made its impact. An added pressure favouring the displacement of the traditional nation state by trans-national associations is the spread of the nuclear weapons technology, paralleling a similar series of events a few hundred years ago when feudal states fell under the impact of gunpowder.

We may expect, therefore, to see a closing of the cycle. Political decisions made over a 20-year period, starting in the 1940s, generated a new technological base which has changed society so much that the political institutions which spawned these decisions, themselves are becoming drastically altered. It is the new technological requirements of the communicative society which are forcing a new set of political systems on global society.

History makes it clear that man's ability as a technological inventor frequently outstrips his capacity to use his inventions wisely. Thus a critical role necessarily devolves upon the political order, which might properly be described as a master technology charged with guaranteeing that technological achievements are absorbed in society in ways that contribute to its stability and general well-being.

Professor R. J. Forbes
formerly of the Department of History of Science and Technology, University of Amsterdam

From *The Conquest of Nature*, Pelican, 1971, p. 124

Scientific

JACK MEADOWS

Professor of Astronomy and History of Science, University of Leicester

We accept a complex interaction between science and technology as being natural. It was not always so. In the seventeenth century, the Royal Society was held up to ridicule for the inapplicability of its science: this at a time when quite sophisticated technology was already in existence. The growth of science-based technology since then has produced a major change in the nature of industry. It has also had the side-effect of making technology increasingly incomprehensible to the general public.

A.J.M.

Engineers engaged on the U.S. space programme have noted wryly: 'When it works, it's a scientific achievement; when it doesn't, it's an engineering fault'. Science owes a major debt to technology, but confusion between the two types of activity tends to disguise the extent of this debt. Until the latter part of the nineteenth century, technology depended hardly at all on science, whereas science frequently drew on technology for its material. The steam engine was developed by engineers using almost entirely empirical methods: but the science of thermodynamics benefited from a study of its workings. By about a hundred years ago, science had progressed sufficiently far for new science-based technologies to become feasible. The discovery of aniline dye led to the growth of a new chemical industry: experimentation with electromagnetism triggered off a new electrical industry. This interaction between science and technology has continued, and increased, ever since: there is now no area of technology that is not in some way affected by scientific developments.

It is, nevertheless, still a good deal easier to observe how science is influenced by technology than vice versa. This becomes clear if we take the U.S. space programme as an example of technological innovation.

The scientific pay-off is obvious—detailed studies of the Moon, space probes to other planets, observations of the Universe in regions of the electromagnetic spectrum inaccessible to ground-based instruments, and so on. But where in the programme has technology benefited from science? If we trace back the history of the U.S. space effort, we might find its original basis in the launch vehicles that were developed for transporting ICBMs. These, in turn, derived from the German rocket bombs of World War II, and, more distantly, from pioneer work on rockets, such as that in the United States, itself, by Robert Goddard. The line of evolution corresponds to a basic model, constantly modified so as to increase both the size and the power. This is a typical story of technological advance, with little obvious new science involved.

Yet it would be possible to argue, on the contrary, that the development of the entire U.S. space programme has been science-based. The necessary qualification to such a statement would be that the science involved has been available, in principle, for some time. Thus the rocket fuels were selected on the basis of their known chemical and physical properties. This knowledge was certainly derived from science, but much of it dated back to the nineteenth century. Again, the flight guidance of a rocket required a knowledge of mechanics; but this represented an area of science dating back, in part, to the seventeenth century. It seems to be a typical feature of the link between science and technology that the science required by technologists is not usually taken from the research frontiers, but from older, 'textbook' material. This may be contrasted with the scientist's use of new technology—as in the space programme—to produce even newer science. It is possible to discern a cycle of operations: new technology produces new science, and this may, in turn, produce new technology, but only after it has had time to bed down and become fully accepted. For example, scientific study of the liquefaction of gases in the nineteenth century resulted in a twentieth-century technology concerned with the production of low-temperature instrumentation. This subsequently permitted widespread scientific experimentation on the properties of materials at low temperature, which is now leading to the growth of a new technology based on the phenomenon of superconductivity. Clearly, feedback occurs in both directions between science and technology. But, owing to the time-lag, it is difficult to point to a specific scientific result and claim that it has produced a particular technological advance, whereas the scientific advances resulting from a particular piece of technology can usually be identified fairly readily*.

The space programmes, both American and Soviet, underline another major feature of the interaction between science and technology: the linkage is often closest in areas of technology that are most remote from our daily lives. The American space programme is supervised by NASA (National Aeronautics and Space Administration), and aviation has always been an area that is costly in terms of R & D expenditure and that impinges on most peoples' lives only occasionally. We might compare aviation, in this respect, with the building industry and the car industry. The products of each of these are encountered every day, but neither industry has ever had much connection with science. Hence, even when science and technology work in close co-operation, the benefit to the general public may be limited. Clearly it is a question of objectives: the drug industry, for example, affects us all, yet it has a traditional basis in science. Whether, or not, a linkage occurs depends primarily therefore on such matters as the funding available, rather than the 'scientific' or 'unscientific' nature of certain types of technology.

We must note, in conclusion, that the current confusion between what is science and what is technology has a perfectly respectable basis. The distinction between pure and applied science grew up in the nineteenth century, as science became applicable. It has never been absolute, and, indeed, when scientists try to justify their work in terms of its potential applicability, they are, themselves, blurring the boundary. Attempts have been made to distinguish between the two activities in terms of methodology: pure science is subject-oriented; applied science and technology are project-oriented. But as the space programme has shown, pure science can also be project-oriented, and this type of approach is becoming increasingly common. Attempts to draw a rigid line between science and technology have become increasingly irrelevant.

* For a more detailed discussion of the interaction between technology and science, see D. S. L. Cardwell, *Technology, Science and History* (Heinemann, 1972).

Moral

JOHN FRANCIS

Director of the Society, Religion and Technology Project, Church of Scotland Home Board, Edinburgh

Technological advances have always raised ethical questions: from the acceptibility of firearms in the Middle Ages to the acceptibility of the hydrogen bomb today. These moral problems have become more acute as the scope of technology has increased. The difficulty is often not the moral goodness or badness of a particular technological innovation, but its level of desirability in a given situation. Thus the ability of computers to store data and speed communication is generally considered valuable; but what if this facility is used to undermine the privacy of the individual?

A.J.M.

1. The Ambiguous Future and the Christian Hope

During the past 5 years the world has passed through an amazing transition which has undoubtedly conditioned the scientific and technological world view. In 1970 it was possible to envisage a world in which all things were still possible with a seemingly endless spectrum of technological innovation stretching forward into the future. There was a vague hint of a more unified objective in the shadow of the first U.N. Conference on the Human Environment held in Stockholm in June 1972. The symbolic form of the arguments surrounding 'only one Earth' did at least indicate that some governments were prepared to acknowledge the threshold of a new era in which choices relating to the content and direction of technological change might be re-examined.

Unfortunately, this vision was not universally upheld and instead, a more artificial constraint in the form of the Arab oil embargo of October 1973, coupled to a quadruple increase in the world price of crude oil in December 1973, brought the whole world momentarily to a watershed of uncertainty concerning the shape of things to come. That mood has subsequently dissolved, although the economic repercussions will probably take a decade to be absorbed fully. What remains is an awareness that this is a turning-point in the perception of social and economic development and of human expectations, which arises from a new understanding of the limits, dilemmas and uncertainties of a scientifically and technologically shaped society.

Over this same period from 1970 to 1975, the World Council of Churches has been engaged in promoting an active exchange of views between scientists, engineers, economists, planners, ethicists and theologians in the light of this rapidly developing situation. After many hours of turbulent debate and discussion in different parts of the world, a consensus view emerged at a conference held in Bucharest in 1974.(1) It was acknowledged that there must be a convergence towards a just and sustainable society incorporating basic human needs, such as adequate food, clothing, shelter and health services. The Christian hope must be that, in a world that seems likely to remain divided for a multitude of historical and political reasons, the response to these constraints may be achieved without any further bitterness or overt acts of aggression. The technological 'fix' has been successful in advancing material progress, and has also promised greater control over the random events in nature that wreak such havoc in many countries through famine, earthquake and hurricane. While the use and deployment of certain forms of technology has become the gateway to economic security and human satisfaction, there is now a sense of an ending which stems from an accepted diagnosis of the problems facing any society that builds exclusively on technological achievement.

From its beginning, the Christian community expressed its faith in an ethic that emphasises the importance of sharing. People saw God's creation as a home, filled with gifts which made possible the sustenance and enjoyment of life, but these gifts needed to be cultivated and shared. When face-to-face with the sobering reality of the present, it is not reasonable to expect that all humanity can live as the most extravagant have been living. In its place, there is a convincing case that the world can still provide the basis for a long-term sustainable state, if there is the necessary will to achieve international co-operation towards this goal. Of course, the search for imaginative technological solutions will go on, but it will be accompanied by a more discriminating view of the social and environmental impacts linked to this process. A considerable degree of innovation can be anticipated in the critical areas of food and energy production, where supply constraints bite deeply into the human fabric. In the case of food, a straightforward expansion of the production baseline would be more than welcome, but it does not follow that a similar expansionist view should be taken of the race to exploit new sources of energy. It is particularly necessary, when the world stands on the threshold of a major move towards dependency on nuclear power systems, but it is equally true of circumstances arising from exploitation of the U.K. Outer Continental Shelf. The struggle to find and to produce energy is set to become the hallmark of the next decade. The ethical dimensions of this task are never very far beneath the surface. It will be worthwhile examining two aspects of the current debate: (1) the risks and potentialities of the further expansion of nuclear power programmes on an international scale, and (2) the social and environmental constraints on the rapid exploration and economic exploitation of North Sea oil and gas.

2. The Pathfinder Role of Nuclear Energy

It cannot be disputed that there are new technologies which are currently the subject of a bitter and public disagreement. Consideration of public acceptability in terms of risk assessment and environmental impact are beginning to hold sway over those factors that are governed almost exclusively by the investment process. Even within the scientific community, there are those who are repelled by the image of an oppressive exploitative model of technology, which they have helped to foster, but which they believe has become self-defeating in terms of human survival. This is best exemplified by the 'limits' approach [2, 3]. There are undoubtedly many more who would agree that we have entered a new stage where there must be more recognition of the questions which science and technology cannot answer. It is in response to these questions that society at large, working with scientists and technologists, must express its convictions about the direction of further technological innovation, and the risk it is willing, or unwilling, to accept. The use and availability of nuclear power on an international scale is a case in point.

From an operational standpoint, nuclear power plants can be brought within the international standards of radiological safety, largely on the basis of engineering measures. These can either be active, by requiring some intervention on the part of the operator (e.g. the movement of a control rod, or the removal of a defective fuel element) or passive, thereby requiring no action by the operator (e.g. a variety of safety devices, biological shields, locks, etc., would appear in this category). There are a number of problems that confront the design engineer arising out of this situation.

1. Each of these measures can fail, but it should be possible in an advanced technology to assess the finite probability of such a failure.
2. With this knowledge of failure probabilities, it is possible to design a second line of defence.
3. Some elements of redundancy will be incorporated into the safety circuits to allow for the incidence of individual component failure; the integrity of the operational safety systems are considerably enhanced by this procedure.

This method of dealing with risks of the first kind arises from a comprehensive, but almost certainly incomplete, understanding of the laws of nature. However, any practising scientist or engineer will acknowledge that it is impossible to measure the initial and boundary conditions on an operational plant with the completeness and accuracy necessary for a fully deterministic prediction of the performance of any technical component. Consequently, a risk of the second kind emerges which cannot be treated in the manner already described, and this introduces the particular dilemma of nuclear power plants, or indeed any other centralised installation dealing in high technology for that matter. The traditional engineering approach to eliminating risks of the second kind has been trial and error. The assessment of risk in the case of these installations can be summarised as follows:

A. Reactor safety in any absolute sense cannot be evaluated by trial and error.
B. Subdividing the problem, i.e. distributing the incidence of risk by minimising the rate of component failure, can lead only to an approximation to ultimate safety.

C. The risk can be made smaller than any small, but predetermined, number which is larger than zero.

D. This remaining 'residual risk' opens the door into the domain of hypotheticality.[4]

This is a difficult point for the non-technical person to appreciate fully, but a great many ethical choices with long-term implications hang on our interpretation of the concept of 'residual risk'. Since there is no prospect of creating a risk-free environment, most people are obliged to evaluate their individual exposure to risk; air travellers and rock climbers alike. It is sobering to admit that most accidents occur in the home, which is merely a reflection of total people multiplied by total time spent in one comparatively safe place. In the present illustration of the finite risk associated with nuclear power systems, the challenge to the scientist and engineer should be more generally understood:[5]

> The process of iteration between theory and experiment which leads to truth in its traditional sense is no longer possible. Such truth can no longer be fully experienced. This means that arguments in the hypothetical domain necessarily and ultimately remain inconclusive ... this ultimate inconclusiveness which is inherent in our task explains, to some extent, the peculiarities of the public debate on nuclear reactor safety. The strange and often unreal features of that debate ... are connected with the 'hypotheticality' of the domain below the level of the residual risk. ... It is impossible (to apply the method of trial and error to ultimate reactor safety) because the consequences of doing so would be too far-reaching. Every country is too small for that—eventually even the globe is itself too small.

This statement by a leading German nuclear scientist suggests that these circumstances are not peculiar to nuclear power, and that the threshold of hypotheticality we are now crossing will characterise the next stages of human enterprise. In other words, it suggests that the magnitude of technological enterprises will be so great that it will not be possible to proceed with absolute certainty that there will be *NO* negative consequences. For that reason, several writers have categorised the utilisation of nuclear energy as a Faustian bargain, although it will be remembered that in Goethe's version, the devil was cheated and Faust went to heaven.

Of course, in the case of nuclear energy, society at large is not faced with such a bargain, but rather with a decision regarding a powerful instrument that can be used or misused. The prospect of a sustainable society is entirely conditioned by our ability to resolve this type of dilemma in a reasonable manner. The question remains 'Can we fully use this period of new enlightenment for the discriminating choices—as in the case of nuclear energy—between infinite potential and infinite risk?' There is very little doubt that plans concerning the future role of nuclear energy will establish an important precedent that may govern the use and widespread availability of other forms of high technology. It is a difficult and often treacherous pathway with no guarantees of a happy or successful conclusion; as individuals we should do far more to acquaint ourselves with the facts of the situation.

3. The Impact of New Technology on Community Life

Whether it is the consideration of a new site for the construction of a nuclear power station close to a major population centre, or a vast tanker terminal on some remote Scottish headland, there is almost inevitably going to be a major disturbance in social and environmental terms for people in the communities most directly challenged by the development. The location of industry, if determined purely by economic considerations, would be governed by such factors as availability of raw materials and energy; easy access to transport facilities, ports and markets. These magnet forces tend to create the conurbation, or megalopolis, and imply social mobility and rootlessness—people moving to jobs, not jobs to people.

Both in Britain and throughout the EEC, however, regional policies are evolving, designed to arrest the decay and depopulation of whole areas and the break-up of communities. But minerals and fossil fuels are distributed by nature, not by man, and their discovery and exploitation may involve the creation of large communities where none previously existed, or the swamping of existing communities by floods of incomers and economic activity. This is the current experience of the north of Scotland, occasioned by the discovery of oil and natural gas in the North Sea.[6, 7]

Few situations could be as dramatic and intense as the Scottish scene today, but all those concerned that economic development should be planned for the advantage of man and his environment should take the lessons to heart. It is possible to suggest a checklist of the major factors that should take precedence.

3.1. *Planning and Local Government*

Despite the objective of planned growth, local authorities can be overtaken by events because:

Before land can be allocated according to its social utility, the speculators move in and the price rises astronomically.

The building industry finds work for incoming industry more lucrative than council house building—so targets are not met, or the cost is too high for local authority budgets.

The typical staff in a local planning department are out of their depth in dealing with experts in advanced technologies. In any event, the technology is moving so fast that the planners cannot keep pace. Instead of a series of contingency plans linked to changes in technology, with recognised social and environmental constraints, there may come a succession of changes of direction and contradictions which leave the community bewildered and insecure.

The commercial operator moves with drive, energy and speed to complete the construction phase and become operational; whereas the response from the public sector is too often slow and uncertain, and local authorities lose out in competition for scarce resources—especially manpower.

3.2. *The New Industrial Environment*

It is well known that manpower planning is an inexact science, but forecasting future levels of employment must nevertheless be attempted in order to make adequate provision for housing and community services. It is common knowledge, too, that demand for labour will be at peak during the construction period of a new enterprise, and it is at this stage, especially if housing is scarce, that established industries and services are at risk because:

(i) employees are attracted away by higher pay;

(ii) internal training and promotion policies are disrupted by labour turnover;

(iii) traditional relativities of job status and pay are disturbed.

In these rather vulnerable areas of the regional economy, if booms and slumps in employment are to be minimised, the long-term survival of existing industries needs somehow to be preserved.

3.3. *Community Reaction*

Most people, not unnaturally, welcome the idea of being better off. But the price they pay is:

a widening gap between rich and poor and feelings of envy towards the possessors of new wealth;

inflated prices in the shops and especially in house prices, leading to indebtedness and then pressure for yet more money;

high level of expenditure on alcohol, gambling, etc.;

the existing community, its culture and way of life, swamped by the cosmopolitan incomers.

Additionally, the community and social services very quickly become inadequate under these new pressures. This is inevitably reflected in the overloading of medical practices, a lack of facilities for young people and especially of voluntary youth leaders, and generally insufficient investment in social and recreational amenities.

3.4. *The Cost–Benefit Matrix*

Based on an evaluation of the planning response to these new aspects of the Scottish scene, there is in evidence more concern in the current situation over sacrificing the physical environment than for the social environment and spiritual values. In those parts of the country most affected by oil-related developments, people are adopting an individualistic approach to life, seeking their security in money and material possessions, and are thereby showing a diminished sense of responsibility towards groups, the community and society.

This suggests that there is a need to restore the balance by introducing a far greater degree of precision into the forward planning machinery at a community level, with the object of stabilising the local economy and providing a proper degree of investment in the community. A vehicle for this purpose might be a 'project group' operating in the community interest.[(8)] Through their participation, local industrialists could contribute to it expertise and management skills. This would be an opportunity for industry and the community to move towards the collaborative management of manpower and resources at a community level. Even if this innovative solution, which would tend to mirror the circumstances of New Town management, cannot be implemented in every location, there are certain lines of action that do apply in the general case:

The schools should make a major contribution to the development of responsible attitudes by teaching that it is the citizen's concern to use the world's resources with the needs of future generations in mind; to make creative use of leisure; to acknowledge responsibility for relevant groups in their society; to look for the intrinsic value of jobs.

Planners should provide a constant flow of authoritative information to the community.

Incoming industry should agree arrangements with local authorities for the construction firms to build houses as well as power stations, fabrication yards and harbours.

There is therefore in this kind of listing the connection between responsible stewardship at the level of general education and the ethical choice facing new industries with probably an undeveloped sense of community responsibility. The struggle to bring these forces into equilibrium will obviously have a major bearing on the future of the regional economy. There is every reason to suppose that once the lessons emerging from this situation are understood, the process of reconciliation can begin in earnest. In Scotland, the future is currently being conditioned by a number of movements in this direction. The moral dimension in the social and political decision-making procedures may thus be reinstated by degrees.

References

1. Science and Technology for Human Development, *Anticipation*, **19**, 1974, WCC, Geneva.
2. D. H. Meadows *et al.*, *The Limits to Growth: A Report for the Club of Rome's project on The Predicament of Mankind*, Universe Books, New York, 1972.
3. M. Mesarovic and E. Pestel, *Mankind at the Turning Point: A second Report to the Club of Rome*, Hutchison, 1975.
4. W. Hafele, *Minerva*, July 1974.
5. Facing up to nuclear power, *Anticipation*, **20**, 1975, WCC, Geneva.

6. J. M. Francis and N. Swan, *Scotland in Turmoil:* A social and environmental assessment of the impact of North Sea oil and gas on communities in the North of Scotland. St. Andrew's Press, Edinburgh, 1973.
7. J. M. Francis and N. Swan, *Scottish Oil Shakedown:* An assessment of some future prospects for oil-related industry in the West of Scotland. Church of Scotland Home Board, 1975.
8. J. M. Francis and J. Swan, *Scotland's Pipedream:* A study of the growth of Peterhead in response to the demands of the North Sea oil and gas industry. Church of Scotland Home Board, 1974.

We may recall that the British Association meeting of 1860 *in Oxford was the scene of the famous confrontation between Huxley and Bishop Wilberforce, between science and orthodoxy. The battle lines have separated, but the mind of man is adrift and the peoples of the civilised world derive their satisfaction from activities which are so often alien to, and destructive both of the physical and intellectual environment. One world view has been eroded and the inadequacy of its substitute is being demonstrated. We have deluded ourselves that through science we find the only avenue to true understanding about nature and the universe. Furthermore, we have persuaded the society in which we work to support our activities in the belief that our discoveries will inevitably, in some way, be of practical benefit. The simple belief in automatic material progress by means of scientific discovery and application is a tragic myth of our age. Science is a powerful and vital human activity—but this confusion of thought and motive is bewildering to man, and it is a most alarming thought that the present antagonisms of society to scientific activity may deepen further.*

Sir Bernard Lovell
Director of Nuffield Radio Astronomy Laboratories and
Professor of Radio Astronomy, Manchester University

From In the Centre of Immensities, Presidential Address to the 137th meeting of the British Association for the Advancement of Science, *The Advancement of Science*, New Issue, No. 1. 29 August 1975

Aesthetic

NIGEL DAVY MOAKES

Publicity Executive and Industrial Designer, Pegler Hattersley Ltd.

That the growth of technology has had a major impact on our aesthetic sensibilities can hardly be doubted; but it is dubious whether any balanced assessment of the situation can yet be made. Clear-cut differences exist, not only between individual opinions at a given time, but between generally received opinion in successive epochs. So Wordsworth in 1833 *emphasised the 'harsh features', of the railway, 'that beauty may disown'; whereas Whitman in* 1876 *praised the 'fierce-throated beauty'. Perhaps the best way of reconciling these conflicting views is by providing informed comment on particular aesthetic changes induced by technological innovation.*

Technology can affect aesthetic sensibility in three ways: by the development of new materials; by producing new devices, which provide subject matter for conscious aesthetic examination; by impinging on everyday life and thought, often via *the unconscious. In one sense, this list is set out in ascending order of importance—with the impact of technology on everyday life both the most operative and the most significant. But it is often impossible to separate these effects from each other. For example, the introduction of concrete as a new material has very evidently led to the development of new styles of building, the appearance of which has evoked plenty of explicit comment and discussion. Equally, the total aesthetic experience of living in a concrete building differs from that of inhabiting buildings created from other materials: not only in obvious aspects, such as difference in scale and layout, but even in such minor points as the method of attaching objects to walls.*

The interaction of technology and aesthetics has come more sharply into focus in the present century owing to the accelerating rate of innovation, but changes in our sensibilities resulting from technological advances have occurred over many more centuries. The English landscape of today, for example, has lost much of the woodland that covered it a few centuries ago—partly through agricultural development, partly through the demand for wood as a fuel and a building material. Our present image of the countryside has thus been defined, in part, by earlier technologies.

No major innovation in technology can occur without affecting the way we view the world, and yet very few innovations initially take any account of our aesthetic sensibilities. There is a fundamental problem for all who are connected with technology, and especially for those in the area of industrial design. It is true, as one designer has said, that: 'There are professions more harmful than industrial design, but only a few of them' (V. Papanek, Design for the Real World)? *Or has industrial design expanded the limits of our taste?* A.J.M.

Modern technology is rapidly changing the appearance of our physical world. The architecture wherein we work and live, the landscape around us, the vehicles which transport us and the everyday articles we use are shaped by engineers, who have a duty to society to develop an intuitive sense of good design in order to contribute to human experience and endeavour.

William Morris (1834-96) saw design in a sociological context; he reacted against the squalor imposed by the Industrial Revolution, against ugly machine-made products, and advocated a return to hand craftwork. Morris saw that skilled craftmanship had been replaced by mechanical routine in apalling working and living conditions. His Art and Crafts Movement did little in his day to improve the design of machine-made objects, which usually remained crude, vulgar and overloaded with ornament. His work did, however, represent a tremendous advance in aesthetic sensibility, which was further developed by Art Nouveau designers. One such designer was Charles Rennie Mackintosh (1868-1928) whose Glasgow School of Art building (1896) is an example of good design (based on a low budget) which attracts by its economy of structure and restraint in ornament.

Lewis F. Day (1845-1910) and John D. Sedding (1837-91) acknowledged that machine manufacture could not be based on handcraft designs, but it was not until about 1894 that van de Velde gave expression to the new thoughts about design in the machine age—'a logical structure of products, uncompromising logic in the use of materials, proud and frank exhibition of working processes' (*Pioneers of Modern Design* by Pevsner). Frank Lloyd Wright (1869-1959) said in 1901: 'Today we have a scientist or an inventor in the place of a Dante or a Shakespeare.' He considered that the machine, 'has noble possibilities'. The lead in the new thinking was taken in Germany where Muthesius (1861-1927), who had spent some time studying domestic architecture and Art Noveau in England, became the acknowledged leader of a drive towards a 'Maschinenstil'. He campaigned against the imitation of hackneyed forms of bygone times. The synthesis of the theory was the work of Walter Gropius (born 1883), whose name is synonymous with the Bauhaus founded in 1919. Gropius regards himself as the follower of Morris and van de Velde. The Bauhaus was an Art school combined with a technical workshop, where architects, designers, artists and craftsmen studied together with a common purpose to discover the possibilities of the machine as a craftsman's tool. The principles laid down by the Bauhaus are the foundation of our approach to Industrial Design today. Since the engineer is largely responsible for the form of the civilised world, it is essential that he should be aware of the human response to his work. Aesthetics should be an integral part of any engineering design.

There is no magic formula and no assured way to create an aesthetically pleasing design. However, there are certain basic elements which can be studied, such as line, space, mass, balance, proportion, contrast and colour. The application of ornament can be in direct conflict with the idea conveyed by function, form and utility. Styling and ornament applied to a design after the function and form have been developed are almost bound to be false. Industrial Design encompasses many aspects, including designed service life, performance characteristics, temperature and pressure ratings, vibration and fatigue limitations, service and maintenance requirements, etc., all of which are considered as a part of 'function'. These various aspects have to be combined with a suitability for the available production facilities, and a concern for cost, as well as aesthetics. It follows that the more restraining the disciplines, the more inventive the designer needs to be. The Design Engineer is often under pressure for a quick and low-cost solution, and, where a profit motive is involved, there is a great temptation for some manufacturers to overlook their responsibilities to society by proliferating bad design. This is in effect a form of pollution.

Conservation is another ideal pursued today by those who are aware that the natural beauty of landscape has been destroyed over the centuries by men greedy to take away natural assets, regardless of the ugliness and chaos they leave behind. Factories, mills, mines, foundries, works, power stations, shipyards and industrial plant cover much of our land, and give employment to mankind. Operating automatic production machinery often gives little satisfaction, since no skill is involved save that of maintaining speed, and working conditions can be unpleasant. These workers may well become insensitive to their environment. The advent of remote control by electronics does at least promise an improvement in working conditions, but will man's intellect and aesthetic sensibility become atrophied?

It is in the field of high technology—in the construction of North Sea oil rigs, Concorde and in space flight—that the computer will play a significant part in fashioning man's environment in the future. The elegant design of Concorde is due not to preconceived ideas, or the study of birds, but to computer analysis of masses of data relating to stresses set up in materials when flying at supersonic speeds in a host of different climatic conditions. The design of the oil rig, similarly, is a brilliant computerised achievement, accounting for immense variations in the most adverse marine environment. On the other hand, power stations and oil rigs are often seen as a blight on the landscape, and the noise of Concorde is a health hazard to those who live within range. In order to survive in this world of diminishing resources and increasing population, we must be prepared to receive new ideas and welcome the potentialities of new technologies, but we must also be intensely concerned with the negative consequences, and with what we may destroy in the name of 'progress'.

For many of us in the Western world colour television, a modern electronic wonder, has become an essential part of everyday life providing relaxation, education and stimulation of the imagination. This may in part compensate and be a substitution for the visual pollution around us, but may also provide the impetus we need to encourage our efforts to enrich our lives by creative pursuits. Today there is a tremendous renewal of interest in all kinds of craftwork, and attendance at evening classes for ceramics, embroidery, woodwork and other subjects is very high. There is universally a great deal of pride in the home, and an active 'Do It Yourself' movement encouraged by manufacturers of well-designed goods for furnishing and decoration.

The world about us, in spite of all the highly sophisticated technology, is still largely very ugly and uncomfortable. The solution to our difficulties lies not in a return to a primitive way of life 'away from it all', but in recognising and responding to new opportunities occasioned by the greater awareness of new social and ecological dimensions. School approaches to science, handicrafts and art are beginning to reflect this trend: greater emphasis is placed on the development of imagination and inventive thought in a social context. The essence of the method is the willingness to encourage the new idea even though it may be in conflict with previous experience. We must learn to make machines our tools and, following precepts of good design, use their great potential to improve our environment and way of life in the most economical way.

Conclusion

The preceeding represent some of the important questions we might wish to ask about the impact on individuals and society of technological innovations. The questions asked, and, indeed, the type of answer sought, depend, in part, on the background and outlook of each questioner. But the overall aim is clearly to place technological advance in the context of the human goals we consider to be important.

One major problem in discussing these questions at the secondary level is that they do not seem to add up to a typical school subject: rather they mix together topics that are normally kept separate in the school curriculum. Hence, even students who are prepared to recognise the relevance and importance of these questions are often not prepared to accept them as a matter for serious study in school. Teachers may find that this situation will change in the next few years as the study of the social impact of technology becomes increasingly common in higher education. Already the basic need for teaching materials can, at least partially, be satisfied. In particular, case studies of technological impact, such as we are considering in this section, are available through such schemes as SISCON (Science in a Social Context). Teachers may find that their use of these materials, besides easing their own teaching problems, will help also to legitimate social studies of technology in the eyes of their students.

PART II

Aspects of Living with Technology

Introduction

There is widespread concern today about the direction our society should take, about the ordering of priorities and the appropriate use of scarce resources in order to maintain and subsequently to improve the quality of life. There is anxiety about trends in technological development, about likely effects upon the health, life-styles and prospects of individuals and societies, and about the widening gulf between developed and developing countries.

In this section, writers comment freely upon a variety of aspects of technological innovation and development: they discuss trends, probable futures, and the implications of choices likely to confront us. What kind of improved husbandry, for instance, can safeguard our dwindling resources? How can we advance the quality of life in our cities? Are we using new methods of transport and communications to best advantage? And the most searching question of all: is man's future of sufficient importance to warrant his urgent concern? Each of these questions provokes different answers according to the degree of emphasis placed on the economic, political, social, individual or other aspects on which we set value. This raises the further questions: whence come our standards of value? What principles should guide us in selecting from the increasing variety of choices presented to us by science and technology?

The overall feeling is one of hope and determination: the common resolve is to achieve a sufficient reorientation of technological effort to ensure improvement in the quality of life on a wider front, and an adaptation of technology to the ways in which people live and want to live. This leads inevitably to discussions of moral purposes and values, and of ways and means in the contexts of Europe and the developing world.

Use of Resources

ENERGY

MEREDITH THRING

Professor of Mechanical Engineering and Head of Department,
Queen Mary College, University of London

A man's food provides him with about 100 W of energy (0.1 kW) and this is equivalent to about 0.12 ton of coal per year. The average energy consumption per person for the whole world is about 1.8 tons of coal/year, i.e. 15 times his food energy: this is often expressed by saying that each man has 15 slaves working for him but much of this energy is in fact wasted by inefficiency and extravagance.

However, the variations between nations in the average energy consumption are enormous. In the U.S.A. the figure is 10–12 tons of coal equivalent per person per year (TCE for short) and in Britain 5–6 TCE. In the developing countries which have two-thirds of the world's population it is only $\frac{1}{4}$ TCE and all these people have only 20 per cent of the total world energy consumption.

Thus the first exciting problem for the next generation of engineers is how to give the people in the under-developed countries ($2\frac{1}{2} \times 10^9$ now and 5×10^9 in A.D. 2010) all the benefits of the Industrial Revolution without the disadvantages such as air, land and water pollution, noise and stress.

The fuel resources of the world can be divided into two categories: fossil fuels which represent either the stored energy of the sun from past ages or the nuclear mass energy of elements in the earth's crust, and renewable sources such as solar energy and wind and water power. Table 1 gives King Hubbert's† estimate of the world's fossil fuel energy resources. Table 2 gives comparable estimates of British energy resources.

† *The Energy Resources of the Earth*, Energy & Power, publ. Scientific American, p. 36.

TABLE 1. World Energy Resources in MTCE
(million tons coal equivalent)

	MTCE
Coal and lignite	7.6×10^6
Natural gas	0.4×10^6
Petroleum liquids	0.44×10^6
Tars and oil	0.066×10^6
Shale oil	0.041×10^6

Approximate conversion factors
1 TCE $\sim \frac{2}{3}$ ton oil $\sim$ 5 bbl oil
$\sim$ 7500 kWh
$\sim$ 2800 ton f-km

TABLE 2. British Energy Resources

	No of years	Coal equivalent Annual rate MTCE/yr	Total reserves MTCE
1. Coal reserves estimated in 1953	250	200	50,000
2. Coal reserves planned for human mining 1975	100	120	12,000
3. Total coal reserves	330	300	100,000
4. Nuclear fission	50	10–30	500–1500
5. North Sea oil	25–40	150–250	1500–5000
6. North Sea gas	25	30–40	750–1000
7. Hydro power	50–100	2	100–200
8. Tidal (Severn Barrage)	∞ (say 300)	1–2	300–600
9. Solar energy	∞ (say 300)	20–50	6000–15,000
10. Wind	∞ (say 300)	1–2	300–600
11. Incineration	50	30	1500
12. Methane from anaerobic fermentation (vegetable, animal and human waste)	∞ (say 300)	5	1500

British coal reserves are thus $\frac{1}{150}$ of the world coal reserves but could, nevertheless, supply our present total energy usage for 150–200 years. Similarly, North Sea oil is in the range $\frac{1}{300}$–$\frac{1}{100}$ of the world's oil reserves. These two tables lead to two clear conclusions for the future use of energy. The first is that we must find a way of bringing coal to the surface as easily as we can bring oil or gas without men going underground.

Coal is the only source of energy which can give all the 7000 million people who will be on this earth in the next century the machinery which can enable them to have a good diet, decent homes, clothes, education and medical services. If we can get the coal out without men going underground then the amount of coal both in Britain and the world will be several times as much as the figures in the tables because the machines can go much deeper than men, work far under the sea and work very thin seams.

The Russians have controlled a pair of hands on the moon by radio from the Earth to take samples of moonrock so we can certainly develop a means whereby the skill of a coalminer can be exercised by him in a mine while he is sitting in a safe clean air cabin on the surface.

The second conclusion is that oil and natural gas should be treated as premium fuels and used only for purposes for which they are uniquely suitable. In the case of oil, this is for road, sea and air transport for which there is no substitute in sight because 1 kg of liquid fuel can be burnt with 14 hg of air which can be collected as the vehicle moves. Oil is also the ideal feedstock for a vast range of petrochemicals. Natural gas should be reserved for small multiple burners, e.g. domestic cooking and heating and small industrial furnaces.

The only-large scale use of renewable energy at present is hydro-power and even for this, one requires a large head of water and enormous flow quantities since 1 ton of water falling 1 km produces only as much heat energy as $\frac{1}{3}$ hg of coal. Thus, hydro-power can only be obtained when there is a large rainfall in high mountains and even then, the capital cost of building a dam and installing the generator turbines is very high. In Britain, most of the available hydro-power is already being used and in the rest of the world the untapped mountain ranges are mostly far from civilisation. Hydropower does, however, provide the only way of storing energy in large quantities as one can run the turbo generators as electric pumps and pump water up into the reservoir with quite good efficiency. The most important problem to be solved with hydro-power is how to stop the reservoirs from being filled up with silt. Both wave power and tidal power suffer even more from the very high capital cost required because of the poor conversion factor and also require to be combined with a storage system or a thermal station which can generate efficiently with big load fluctuations.

Wind power can produce about 120 W/m^2 in a normal breeze so that to produce 3 kW for a house, one would need 25 m^2 of blades or a mill of diameter 5.6 m and even this would only provide current when the wind was blowing.

Some very interesting work is being done on vertical axis windmills which do not need a high tower to support the axis and do not have to be turned into the wind. There are, however, very interesting and fuel-saving possibilities in the heating of water in a large tank by a combination of a solar heater and a windmill-operated heat pump. Solar heating can provide up to 1 kW/m^2 as low-grade heat, temperatures of 45°C being obtainable. Both solar heaters and windmills are expensive in first installation but once installed can function almost indefinitely so that we must develop and install these all over the world before our fossil fuels run out. The other exciting possibility being studied is to find a series of solar-operated chemical reactions which produce a combustible fuel such as hydrogen, methane and some other hydrocarbon or alcohol from H_2O and CO_2. This would be speeding up and rendering more efficient Nature's growth processes.

Geothermal energy is in theory a free source of energy but in practice it is extremely difficult to obtain the necessary heat transfer through the rocks to produce steam. Nuclear fission is extremely limited in its potential to supply electricity because the construction of the power stations with all the necessary safety precautions and waste-disposal methods absorbs so much of our limited resources and we cannot afford to cover the country with burnt-out power stations which can never be touched. Moreover, the greatest way to save energy is to use small power stations near to factories and homes which can use the low-grade heat left after generating electricity and nuclear power stations must be remote from inhabited areas.

This leads to the final conclusion. Since fossil fuels are limited, since world population will double and since it is impossible to have a permanently stable world with gross wastage in some countries and shortage in others, it follows that the rich countries will have to learn to live a good life with an energy consumption *per capita* only a fraction of what it is now. This presents the most exciting challenge of all to the engineer: how to grow enough good food and provide travel and transport as good as it is now, and all the worthwhile consequences of the Industrial Revolution, on about the present world average energy consumption of less than 2 tons coal equivalent per heat per year.

MATERIALS SCIENCE

JAMES GORDON

Professor of Materials Technology, Department of Engineering and Cybernetics, University of Reading

There is much evidence that Faraday was intensely interested in the ordinary behaviour of common objects and familiar materials but he had the greatness to realise and to accept the limitations of the science of his day. His trouble, as he repeatedly and wistfully recognised, was that contemporary molecular science was not adequate to tackle such subjects.

In fact, electricity—and indeed astronomy—is a much *easier* proposition, experimentally and conceptually, than the bending of metals or splitting of a piece of wood. The conceptual difficulties are recognisable as soon as we begin to think about the poverty of the vocabulary with which we attempt to describe the way in which ordinary solids deflect and break. What do we really mean by 'strong' and 'stiff' or 'tough' and 'brittle'? Is 'strong' the same thing as 'tough'? Is 'weak' the same thing as 'brittle'? As for the experimental difficulties, I would recommend the sceptic to try to determine, with an ordinary microscope, why thin glass fibres are about a hundred times as strong as ordinary glass. It can be done but it is not easy.

In fact, the whole problem of why a piece of glass will break when a piece of copper will bend was just too difficult for the Victorians and indeed for scientists down to very recent times. Rather naturally, therefore, lesser mortals than Faraday, who were puffed up with steam turbines and X-rays and astronomy and the wonderful invention of wireless, most generally took the view that problems about the behaviour of ordinary solids were too trivial and unimportant to attract the attention of serious men of science. Some smell of this attitude still hangs around the classrooms today.

Metals like steel are important, technically and economically and politically. But why? What are the properties which make steel so useful? How are we to define these properties? How is it that metals, and metals alone, possess these properties? In fact, of course, it is all due to something called the 'dislocation mechanism'. But dislocations in crystals have only been discovered fairly recently and the story is one which combines a certain amount of drama with a great deal of sheer intellectual satisfaction. In a rather similar way, wood is immensely important, both in nature and in technology. Its mechanical properties are quite exceptionally good. Weight for weight, they are rather better than steel. The reasons for this, the fine-scale mechanisms which make the behaviour of wood totally different from that of the sugars from which it is made, have only been discovered very recently indeed. In fact, within the last year or two. Again, the story is one of almost incredible subtlety and elegance. Almost daily, modern materials science is revealing the fantastic beauty and cleverness of the mechanisms which enable the common materials in our lives to do their simple, commonplace jobs. It is just the materials which we are apt to regard as too ordinary to be worth our attention which seem to perform in the most interesting ways.

The appreciation of things like this is, no doubt, good for all of us, if only to stimulate the senses of wonder and humility. In addition, we really ought to know how things like steel and concrete and paper are made: just as we ought to know where Japan or Madagascar are to be found on the map. Women may be the worst offenders here. How many girls have the faintest idea what modern shoemaking machinery is like? What are the properties of different plastics? How is nylon made? What does collagen break down to when we cook meat? How are shear stresses and Poisson's ratio relevant to dressmaking?

The importance of elasticity and materials science in technology is so obvious. The trouble in the past has been that the simple problems have just been too difficult. Now that door is unlocked and we are really beginning to understand what makes materials. We are also beginning to understand how to make entirely new materials that have never existed before. How far we are wise in what we do here is an interesting point and another question, but one which people will have to face before very long.

Until quite recently most doctors and biologists have resisted all engineering ideas about materials and elasticity with all the force of their emotions. From time to time I get letters telling me that it is futile, and probably morally wrong, to apply the ideas of structural engineering to biology. But nature uses methods of immense subtlety in her problems of chemistry and in her control mechanisms and it is really absurd to suppose that her approach to structures and elasticity would be a crude one.

Every plant and every animal has to resist mechanical loads of one kind or another and mechanical failure very often means death. In fact mechanical failure is rather rare in animals which most often die from other causes. This does not mean that mechanical performance is unimportant; it means that nature has been exceptionally good at solving the engineering problems with which animals are faced. Naturally, it is highly instructive to see how nature manages to do it. The science of biomechanics is still at rather an early stage but we are already beginning to get some idea of the way in which animals are able to carry loads with safety. Once again, the emotion one is left with is an almost numinous sense of wonder.

Dyed-in-the-wool professional engineers of the old-fashioned sort are not very interested in the structural problems of trees and worms and kangaroos, but this is stupid and rather arrogant of them. All the indications are that there is a convergence and interaction between biomechanics and technology. Traditional metallic nut and bolt, girder and cogwheel and steam-engine technology is barely 200 years old. There is nothing especially holy about it and many of us would be quite glad to see it go. This sort of engineering is almost wholly dependent on cheap and plentiful supplies of energy—which may not last much longer.

But nature has always been short of energy. Basically, plants, such as trees, have to stretch out to collect the scarce and fitful energy of the sun. To do so they need a structure which is both diffuse and also cost-effective. This is just what the nut and bolt engineers are so bad at and what nature is so good at. We can learn a great deal from trees, if we are not too proud to do so. Again, weight for weight, the strain energy storage in the tendons of a kangaroo or a squirrel, or even a skier, is about fifteen times as high as the best modern steel spring. The Greeks realised this when they made their very efficient catapults with ox-tendon springs. Yet again, steel super-tankers and steel bridges break in two with disastrous results, so do aeroplanes, if they are given the chance. Even rubber balloons go pop if they are pricked with a pin. The soft tissues of healthy animals *never* behave like this. In short we have an awful lot to learn and the sooner we start doing so the better.

WATER

JOHN McIVER

Reader in Microbiology and Director of the Water Resources Unit, Trent Polytechnic

It is little appreciated, but nonetheless true, that the maintenance of our modern life style is as dependent on the successful operation of our effluent treatment plants as it is on successful trading balances or the availability of raw materials including the ubiquitous petrochemical. To assess the validity of this, one need only examine the mortality records for the past four centuries. From a norm of 100–500 deaths per thousand population (neglecting the influence of the great pandemics) we have today arrived at a position of restabilized population growth based on low birth and death rates, each in the region of 20 per thousand population per annum. While it would be ridiculous to attribute this improvement in public health solely to developments in water and effluent quality, it is true to say that it can be more closely correlated to the care and attention of public health engineers than to the early ministrations of the medical profession. To expand on this premise, may I point out that, although we all have a fairly clear idea of the sociological causes for the appearance and spread of diseases causing those early high death rates, we are not at all clear, scientifically, why they disappeared. It certainly was not due to a development within the population of immunity, nor to the application of noxious medicines and barbaric surgical practices. No, the real answer must lie in an empirical appreciation of the causes and mechanisms of disease dissemination, even though the agents themselves were as yet unidentified, and to the breaking of this chain of events through admittedly imperfect waste treatment and the judicious selection of potable water sources.

I have drawn your attention to this most important aspect because we will increasingly become aware of the fact that the U.K. is short of potable water, although admittedly now restricted to particular times of the year. There is, also, no reason to believe that the problem will go away of its own accord; indeed, as both the population and its demands for consumer goods arise, so will its demands for water. This presents us then with a very real dilemma. Do we lower our highly respected and successful standards with possibly dangerous consequences, or do we financially and technologically meet this need which undoubtedly will cause other perturbations? Possibly some of the considerations now under study will give us a lead.

The first option open to us is to pursue present policies by commissioning further reservoirs, thus evening out seasonal fluctuations through storage of clean water for high-demand periods. This is, however, meeting increasing resistance by, at least, sections of the populace concerned about the loss of agricultural, ecological and aesthetic resources. It is interesting to note in this respect that aspects of this concern are now being circumvented in planning proposals by insistence upon joint amenity use of such water bodies. An extension of this storage idea is to initiate planned barrage schemes which would allow the filling of certain salt-water estuaries with fresh river water, following damming. It is important to note in this context that the water is likely to be, at least, slightly contaminated with agrochemicals and effluents following inland travel to the basin and that extensive piping costs would be involved to transport the necessarily treated water back to urban areas.

A second option arises from the realization that most population centres were, for econo-historical reasons, established on rivers, albeit now polluted to varying degrees. Should it be possible, therefore, to raise the standard of effluent discharge to reduce their load and/or to establish effective treatment processes for water abstraction, this approach would, at least, alleviate transport costs and meet all foreseeable needs. There are, of course, several variants to this theme. One is to continue the policy of treating all domestic and most industrial water to the increasingly rigorous standards necessary for public health supply. This would, however, mean a considerable capital expenditure on new works when it is realized that under various estimates up to 60 per cent of our present effluent treatment plants are now overloaded and discharging partially treated effluent.

It is appreciation of this which raises the matter of dual supply, i.e. a potable water for domestic use and for processes where the water arises or comes in contact with consumables, and a lower grade of water, possibly partially treated for all other industrial needs. Although this is attractive in theory, as it greatly reduces treatment needs and is implementable even now, it does raise certain problems arising from the mechanics of installation and maintenance. It would require the laying of a second distribution system alongside that present with all the resultant disruptions and exorbitant costs. Even more seriously it would raise the possibility of cross-connection and seepage contamination between the two and therefore might not allow a continuing guarantee of public safety from the potable supply—a consideration so serious in its implications that most authorities will not embrace its logical conclusion of *full* dual supply. Thus, it is probably limited to separate protected installations, to specific industries or complexes with high demand for low-grade water.

A third option is to examine industrial and domestic needs in order to reduce their demand for high-quality supply. The most obvious approach here is to raise the cost of water and/or to implement metering of use. Both of these are likely to cause reconsideration of need, and indeed industry is already recognizing water to be an important manufacturing cost, with steps being taken, individually, to reduce this factor. Pursuance of this approach raises further consideration for domestic use where toilet-flushing accounts for 40 per cent of household consumption. This water is only required for solids transport and obviously could be of a lower quality than at present used. Certain highly developed communities are now considering dry closeting to remove this demand entirely. However, where mains drainage and treatment are already established another tack would be to re-use sink, bath and washing machine water for this purpose. Present considerations generally involve a small domestic chemical or physical treatment unit for clarification and microbial removal to protect small delving hands.

What is clear from these deliberations is that there is a real need for an informed populace cognizant of the options open to them and aware of the standards to be met. The water industry has carried out its functions within economic constraints to a remarkable extent.

It will, however, need our assistance to continue this policy.

For Further Reading

1. Annual Reports of the Regional Water Authorities.
2. *Taken for Granted.* Report of the Working Party on Sewage Disposal.
3. *Science and Public Health,* AST 28110, The Open University Press.
4. K. Smith, *Water in Britain,* Macmillan, 1972.
5. Dr. A. L. Downing, (Ed.), *Water Pollution Control Engineering,* HMSO, 1970.

FOOD

DIANA MANNING

Research Fellow in Interdisciplinary Science Studies, Middlesex Polytechnic

There are an enormous number of misconceptions surrounding food and a great many are held, and fiercely guarded, by such respected personnel as doctors, economists and scientists. Here are a bunch of them: perhaps you can think of others.

1. Two-thirds of the world's population is starving.

This claim is based on a mistake made in the 1950s by Boyd Orr, then the director of FAO (the U.N. Food and Agriculture Organisation), who misread some statistic and it has since become a popular cliché.[4] So what *is* the present situation?

The literature is full of diagrams such as the one shown in Fig. 1 (which was taken from the Club of Rome's report: Limits to Growth).[10] From this it would seem that the diet of the average Indian or African is grossly deficient in calories and even more so in protein. Moreover, the

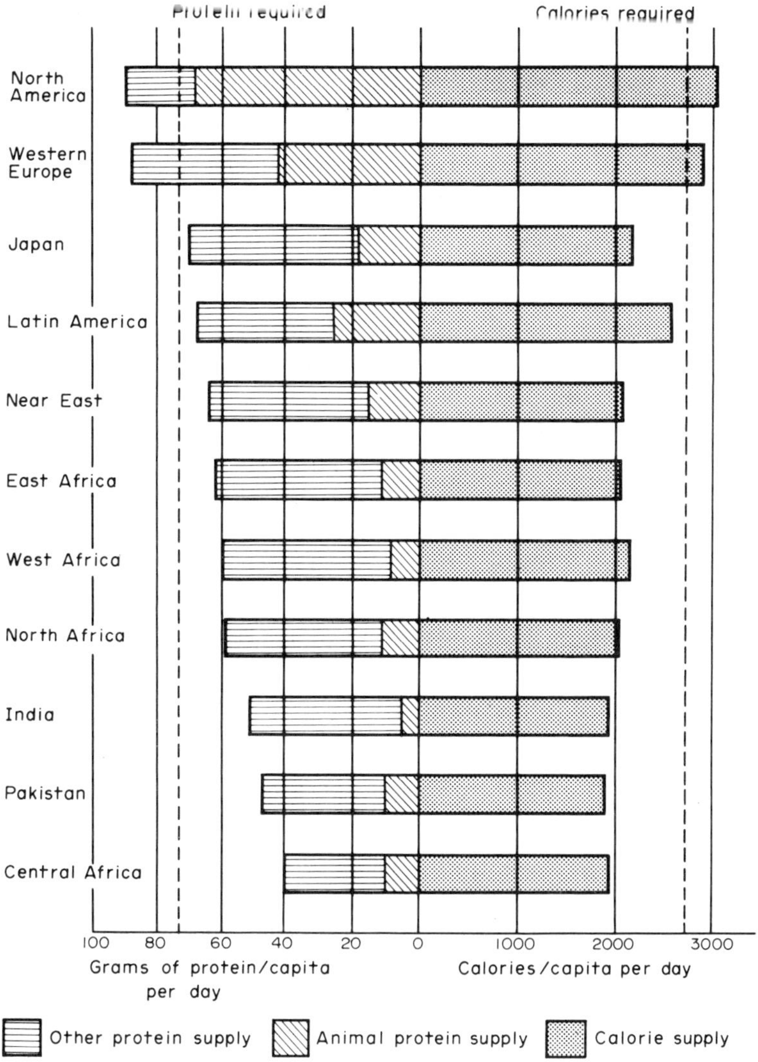
Protein required
Calories required
North America
Western Europe
Japan
Latin America
Near East
East Africa
West Africa
North Africa
India
Pakistan
Central Africa
100
80
60
40
20
0
1000
2000
3000
Grams of protein/capita per day
Calories/capita per day
Other protein supply
Animal protein supply
Calorie supply

Figure 1.

bar charts are based on average figures, so many people must eat even less. If this is true, it would appear to support the 'two-thirds-of-the-world-is-starving' claim. But the fact remains that the vast majority of adults in developing countries manage to maintain body weight, work and have children. Could it be that something is wrong with our estimate of nutritional requirements? It seems that people adapt remarkably well to their diet and it is not always possible to apply 'standards' worked out for affluent Western populations to other communities. However, there are no grounds for complacency. Groups subsisting on these minimal diets are sub-optimally nourished, and their health and ability to work would probably improve with more food. 'Undernutrition' of this type is difficult to assess, but that does not mean that it does not exist. It is certainly true that those who subsist on a low food intake are very vulnerable to any disruption of their food supply and some groups, notably small children, and pregnant and lactating woman, are prone to 'frank' (i.e. clinical) malnutrition. In many parts of the world, 50 per cent of the children die before their fifth birthday, and poor nutrition is largely responsible.

2. The chief cause of malnutrition in young children in the Third World is protein deficiency.

This view has been held by the nutritional establishment for two decades: hence the emphasis on dried skimmed milk and high protein food mixes in food aid and emergency relief. Figure 1 certainly suggests a frightening 'protein gap', but again we find that estimates of protein requirements are not as absolute as they seem. In general, they have been falling steadily since the 1940s (if present trends continue we won't need any protein by the year 2000!) and a recent report by the British Overseas Development Ministry claims that a diet contains a safe level of protein if 6 per cent of the calories come from protein *and the total number of calories is adequate.*[14] By this criterion, most staple cereal foods are adequate in protein providing enough food is consumed. It seems that marasmus and kwashiorkor, the hideous nutritional diseases we see in the Oxfam posters, result from inadequate food intake (usually in combination with other diseases).[6] There are many reasons why children may not eat enough. The food may be too bulky or unpalatable or the child's appetite may be impaired because of other diseases resulting from lack of hygiene, the mother's ignorance, or a host of other environmental factors known collectively as poverty.[16]

3. The world cannot at present produce enough food to feed its population.

Nonsense! We may not be able to support the 7 billion people who will be alive in the year 2000, or the 14 billions who may be around in 2030, but there is no doubt that our present 3.7 billions *could* be adequately fed if resources were equitably distributed.

At present, the amount of grain which is used to feed 200 million Americans could support 1500 million Chinese on the average Chinese diet.[9] That does not of course mean that Americans stuff themselves with rice and bread, but they eat a great deal of meat and intensive livestock production requires high-quality animal feedstuffs: cereals, fishmeal and seedcake. Britain imports 60 per cent of all her food and 50 per cent of her animal feed, much of it from the developing world.[19]

Besides these obscene inequalities between rich and poor countries, there are huge gaps between rich and poor in the same country. Many rural societies have an effectively feudal land tenure system and the poor must forfeit sometimes as much as 70 per cent of the produce of the land they farm as rent to the wealthy landowners.[12]

Anyone who is not a wholly self-sufficient subsistence farmer must have money to buy food. If you cannot get a job or food prices rise to a level you cannot afford, you may starve to death from lack of money, even when there is no absolute food shortage.

4. Malnutrition may be eradicated by the use of modern, sophisticated agricultural technology to improve food production.

This is one of the most popular and widely held misconceptions concerning food and is a reflection of our faith in the power of the 'technological fix' to solve all our problems.[1] The 'Green Revolution' was a term coined by the popular press in the 1960s to describe the development of high-yielding strains of cereal crops and it was heralded as the ultimate solution to world hunger. It is quite true that new strains may permit enormous increases in yield and selective breeding can produce crops more resistant to fluctuations in weather and other environmental pressures. But alas! In this imperfect world it is impossible to get something for nothing and according to the Second Law of Thermodynamics—we cannot even break even! The big snag is that these strains require massive inputs of fertilisers and water and these are very expensive, so the new strains can be used only by the wealthier farmers.[20] For production to be economic, 'monoculture' must be adopted, i.e. large areas of land must be devoted to a single crop. The short-term financial returns from this practice are very great and contribute to the trend towards bigger and wealthier farms. Small farmers are bought out and are forced to become wage labourers, but as the larger farms become more prosperous, human unskilled labour is replaced by oil-dependent agricultural equipment.[13] Those made redundant often drift to the towns and join the ranks of the unemployed in the squatting settlements which border major cities in all developing countries. 'Well that's tough,' you may think, 'but the increased agricultural production contributes to economic growth and must indirectly improve prosperity for all.' Even if that were so, unfortunately the increased yields are not always permanent.

The stability of an ecosystem is determined by the number of interacting species it contains. All agriculture involves an artificial simplification of the local ecosystem, which is why farming is such hard work. Monoculture is an extreme version of this and the system is so unstable that a constant battle must be waged against predators, plant diseases, weeds and soil erosion. The battle is normally fought by means of more expensive, energy-consuming inputs, e.g. chemical pesticides, which may produce more undesirable ecological side-effects.[1]

Britain has the most productive farm land in the world, but three to five times as much energy is used up by our agricultural techniques, etc., as is eventually consumed in the form of the food produced. Furthermore, the energy expenditure comes ultimately from fossil fuels, i.e. energy 'capital', and that which is yielded comes from the Sun, which is equivalent to 'income'.[2,8]

5. More schools are needed in developing countries to educate the people in new and better farming techniques.

Whilst sophisticated technology can enormously improve food production and preservation, it is expensive and can give rise to unforeseeable ecological and cultural side effects, so its use in rural development is a bit of a gamble. At last some research is proceeding into 'appropriate' technologies, i.e. small-scale ecologically compatible, labour-intensive, cheap technology, suitable for poor, rural communities. There is no doubt that changes in agricultural techniques can improve food production and these need not require the complete overthrow of traditional methods.[15,18]

Any change in custom requires education. In most poor countries, formal education is modelled on the Western schooling system. It is academic, specialised, competitive, expensive and culturally alien, with a high drop-out rate at each level.[7] Even primary schooling is seldom available to all and only a tiny minority of privileged pupils get to the top of the educational pyramid. These people usually acquire Western aspirations in the process and tend to seek high-status employment in the modern sector. Though attempts are being made to increase the relevance of the subjects taught (I believe Latin is no longer compulsory in West African schools!), education still does not reach the bulk of the rural population, especially the women. Women determine family health and nutrition in all societies and in many parts of the world, most notably in Africa, they are also responsible for agriculture, especially food production, yet they are scarcely touched by the education system, including the special practical training courses for farmers.[3] Only a small proportion of girls go to school in most poor societies, partly because of sheer prejudice, but also because they have so much work to do. No significant rural development is possible without the full involvement of women. Population control, for instance, will only be accomplished by a revolution in women's status and consciousness. To reach the bulk of the rural population, especially women, education in basic agriculture, hygiene, nutrition and literacy must by-pass the school system and reach the people where they are: at home or at work. This sounds difficult, but non-formal education projects of various types are now in progress in many developing countries and the results are promising.[7]

So far I have talked about the Third World as if the food crisis does not involve us. This is probably the biggest misconception of all.

It is easy to blame our own explosive food price inflation on some convenient scapegoat: the Common Market, the Government, the Trade Unions, etc., but eventually it must penetrate our thick British skins that we cannot continue to support ourselves on the resources of the rest of the world.[5] Unfortunately the laws of economics are counter-intuitive and a sudden reduction in our food imports from poor countries would be damaging to all concerned. However, our excessive demands on world grain reserves could and should be reduced and soon we may be unable to afford our present level of food, fertiliser and energy imports.

Technological innovations such as the use of micro-organisms to fix nitrogen to fertilise our crops, development of single-cell proteins for animal feed, improvement of hillside grazing areas and reclamation of derelict land can increase our self-sufficiency, but by far the most significant factor is diet. We must reduce our consumption of animal products.[11]

Many of these changes involve long-term agricultural policy, but there is also scope for grassroots actions. Home food production, measures to cut waste, organisation of food co-ops and many other projects can be undertaken by anyone equipped with sufficient skills and motivation and these qualities can be encouraged by education, both formal and 'non-formal'.

It is really rather shocking that our present education system does not provide us with an understanding of the basic processes of our own survival. Everybody should be made aware of the interactions between our society and the rest of the world, and between our species and the rest of the Biosphere. But consideration of global issues alone can overwhelm the individual with feelings of futility and personal impotence. It should always be supplemented by some practical project, near at hand, which can yield results, however small.

Opportunities for projects to do with food abound and students of every age group, from primary school upwards, can try out something to do with cooking, cultivation, urban food production, or waste reduction. Some activities could involve the students' families or the local community: just how much food waste is there from the shops, restaurants and households in a particular street? Could it be used as compost or animal feed? How much would it cost to collect and re-use it? How much energy would be required? This type of inquiry could be fitted into existing courses in the sciences, engineering and economics and illustrate their social relevance, and might also contribute to the 'non-formal' education of the lay community.

It is easy to make sweeping recommendations in the certain knowledge that they will not be adopted because the political and economic forces which shape our society are so remote from our control. But that's only half the story. The social revolution which will be necessary to bring about a just society which can survive in a finite world is largely an internal one. There are practical things we can do: skills we can learn and projects we can try out, as individuals or in groups, but more especially we will have to change our attitudes to work, wealth, nature, technology, social roles and human relationships. That will be the most significant part of the revolution and we can start it right now.

References

1. M. Allaby, *Who Will Eat?* Tom Stacey: Prospect for Man Series (1972).
2. K. L. Blaxter, Paper presented at conference on *Chemical and Mechanical Energy Utilisation in Agriculture*, Society of the Chemical Industry, London (October 1974). Can Britain feed herself? *New Scient.* (20 Mar. 1975), pp. 697–702.
3. E. Boserup, *Woman's Role in Economic Development*, George Allen & Unwin Ltd., 1970.
4. C. Clark, In discussion following presentation of paper by P. U. Sukhatme, at the Royal Statistical Society (17 May 1961) on *The World's Hunger: Future Needs and Food Supplies*. Reported in *Proc. Roy. Stat. Soc.* (1961), p. 512.
5. The Ecologist, *Blueprint for Survival*, Penguin Books, 1972.
6. D. M. Hegsted, A deprivation syndrome or protein–calorie malnutrition? *Nutr. Rev.* **30** (1972), pp. 52–54.
7. I. Illich, *Deschooling Society*. Calder & Bovars, 1971.
8. G. Leach, *Energy and Food Production.* Report for the International Institute for Environment & Development, 1975.
9. J. Mayer, Nutritionist, Harvard University.
10. D. H. Meadows, D. L. Meadows, J. Randers and W. W. Behrens, III, *Limits to Growth.* Report for the Club of Rome project on the predicament of mankind. Earth Island Ltd., London, 1972.
11. K. Mellanby, *Can Britain Feed Itself.* Merlin Press: London, 1975.
12. What really happened in Ethiopia. *New Internationalist* No. 10 (Dec. 1973), pp. 8–9.
13. Editorial comment on the Green Revolution. *New Internationalist* no. 24 (Feb. 1975), p. 1.
14. ODA Advisory Committee on Protein, *British Aid and the Relief of Malnutrition.* Ministry of Overseas Development (July 1974).
15. V. Papanek, *Design for the Real World*, Paladin Books, 1974.
16. P. Payne, Protein deficiency or starvation? *New Scient.* **64**, no. 922 (7 Nov 1974), pp. 393–395, 398. In special issue on *The Hungry Planet.*
17. M. Pyke, *Man and Food*, World University Library, 1970.
18. E. F. Schumacher, *Small is Beautiful*, Bland & Briggs Ltd., London, 1973; Sphere Books, 1975.

19. E. Stamp, The end of cheap food. *New Internationalist* no. 2, (April 1973), pp. 6–8.
20. J. Tinker, The Green Revolution is over. *New Scient.* **64**, no. 922 (7 Nov 1974), pp. 388–393. In special issue on *The Hungry Planet*.

Cities

THE DESIGN OF A NEW TOWN

JOHN TAYLOR

Head of Town and Country Planning, Trent Polytechnic

New Towns are the test bed for urban design ideas and Milton Keynes is arguably one of the most exciting planning projects ever undertaken in Europe. In scale it dwarfs all earlier British new towns and its range of innovatory activity far exceeds previous ambitious architectural endeavours. What makes Milton Keynes so different? In short it is an embryo of an 'instant build' city conceived for the future—a town design exercise which hopes to stand the test of time. No effort is being spared to carefully consider the needs of its citizens, its surrounding population and the fruits of modern technology and contemporary thinking. As a consequence it is hoped that by the year AD 2000 perhaps a quarter of a million inhabitants will have come to regard the city as their 'home'.

The site chosen for the city is a 'green-field' site just off the M1 motorway some 70 miles north of London so the planners have not been faced with a major problem of adapting existing urban fabric. The designated area does, however, contain a number of established small communities such as Stoney Stratford, Bletchley, and the village of Milton Keynes itself. To mastermind the shaping of a new environment on this rural site the Board of the Development Corporation commissioned planning consultants Llewellyn Davies, Weeks, Foresteer, Walker and Bor to prepare a series of guidelines. Their consultant's plan published in 1970 aimed to ensure:

(a) opportunity and freedom of choice;
(b) balance and variety;
(c) easy movement and access;

(d) creation of an attractive city;
(e) public awareness and participation;
(f) efficient and imaginative use of resources.

Their proposals sketched out the shape, size and form of the city and, in very broad terms how and where houses, transport, education, commerce, industry, health and leisure would function. The plan was not a rigid one as it was felt that within the thirty or so years it would take to build the city, new technology and fresh thinking must have a place in the development process.

The Development Corporation is now implementing the plan from its headquarters at Wavendon Tower at the eastern edge of the city. Here the Corporation's estates specialists acquire land and negotiate with industrial and commercial organisations while architects design houses, factories, offices and public buildings, and engineers organise the provision of basic services and communications. In essence the Corporation have to co-ordinate the overall development process. They must ensure a balanced provision of key components such as people, houses, employment, transport, shops, schools and the like. Activities must be programmed to maintain a reasonable equilibrium and satisfying environment as work progresses. All this must be achieved with an eye to the needs of the individual and the search for maximising opportunities for a pleasant life in an urban environment.

As Milton Keynes is reaching a stage when residents and visitors alike can see the whole project taking on a coherent shape it is clear that the city is bound to be very different from traditional settlements. The fact that variety and freedom of choice are two fundamental principles governing not only the design of houses but the planning of the whole city is very much apparent. The new city is a 'low-density' development with clear cut traffic arteries, extensive pedestrian areas and a rich blend of some of the best new and traditional ideas in housing design. In this setting it is not difficult to believe that almost anything will be tried, tested and evaluated. Consequently it is no surprise to find a pilot-project solar-heated house not too far away from the 45 acres set aside for woodland nurseries which will implement a massive programme of planting design to make Milton Keynes 'a city of trees'.

By any standard the decision to provide homes for a quarter of a million people in one concerted effort is an imaginative development.

People who come to Milton Keynes in the green heart of Buckinghamshire are bound to find a community which is already different. All inhabitants are facing significant change—those who have lived in the area are seeing their environment dramatically altered; new arrivals are faced with radically new concepts and surroundings. Everyone in this context shares the experience of learning and deciding how the city of the future should work and be planned. On the evidence at hand it is clear that Milton Keynes is already well worth a visit and demands close scrutiny by all concerned with improving our quality of life.

SAFETY AT WORK

KENNETH GREEN

Department of Liberal Studies in Science, University of Manchester

According to the official statistics, 12,197 people were killed as a result of accidents at work in Great Britain from 1961 to 1970. Another 9378 died from certain officially recognised ('prescribed') diseases contracted as a direct result of their work, diseases such as pneumoconiosis, asbestosis or cancer of the bladder. In that same period there were, on average, every year between 450,000 and 550,000 accidents at work which led to the worker involved being absent from work for more than three days. A further 20,000 to 25,000 per year were absent from work due to certified incapacity as a result of prescribed diseases contracted directly from work activities. These figures considerably *understate* the fatalities from and accidents at work. Some accidents which, by law, are supposed to be reported are not reported; some 5 million to 6 million work people (about 20 per cent of the total workforce) did not fall within the scope of the various occupational safety and health regulations in force up to the beginning of 1975, so that accidents to them were not legally notifiable and are, therefore, not necessarily recorded in any of the official statistics; and accidents which cause absence from work of less than three days are not included. So, more realistic figures would be something like this: every year, 2000 killed in accidents at work, 1000 dead as a result of prescribed industrial diseases, 1 million injured or absent from work for at least 3 days due to prescribed diseases, and 10 million injured needing first-aid. Whatever the 'real' figures, it is recorded that at least 23 million working days have been lost annually over the last 15 years on account of industrial injury and disease; this, on average, is well above the number of days lost through industrial disputes.

Deaths and accidents directly from work, large in number though they are, are perhaps not the most worrying feature of the problem of work hazards. It has been estimated that every year, in America, about 3000 new chemicals are introduced into industry. Since scientific argument still rages over the effects on people's health of such long-used substances as lead or zinc or mercury, it is hardly surprising that conclusions have not been reached on the effects of the vast majority of these 3000 chemicals. In addition it is likely that in the future many chemicals which have been in use in industry for some time will be found to have undesired and uninvestigated effects upon the workers who work with them. Recently, for example, there has been concern over the unexpectedly high incidence of liver cancer amongst those workers involved in the production of the widely used plastic polyvinylchloride.

There is now considerably more awareness of the health and safety problems related to work. The Flixborough disaster, press and television attention over such occupational diseases as asbestosis and pneumoconiosis, general public concern over 'environmental' problems, and the reorganisation of the official administrative structures relevant to occupational health and safety matters following on from the Health and Safety at Work Act which came into force at the beginning of 1975, have all brought such health and safety issues into greater prominence.

What contribution can the schools make to develop and strengthen this current increase in awareness? The teaching of health and safety relating to certain activities has been a concern of schools for some considerable time, whether it be lessons in road safety (teaching young children how to cross the road or teaching bicycle-riders the Highway Code), safety in the laboratory, or elementary hygiene as part of Physical Education. Schools have also become closely involved in the campaign against smoking, by attempting to prevent children taking it up in the first place. But the discussion of questions concerning health and safety *at work* has perhaps not been given the attention it deserves in schools. And yet it could be said that such discussion is now almost required by law! The 'Health and Safety at Work Act, 1974' imposes upon employees, as well as on employers, responsibilities for safe working. As the Act puts it (Section 7), 'It shall be the duty of every employee while at work . . . to take reasonable care for the health and safety of himself and of other persons who may be affected by his acts or omissions at work.' An understanding of the possible hazards the school student is likely to face when employed and of how he or she should act in hazardous situations when placed in them at work could very easily begin in the schools. It could certainly be part of the curriculum in Technical Colleges.

What particular things can schools and colleges do to begin such an understanding? Here are a few suggestions of the types of activity around which the subject could be organised:

1. The detection of the levels of potentially dangerous materials in the air of, say, a chemical factory is obviously important if control over such materials is to be effected. Equipment is available which can be used for such monitoring*. Obviously use of such equipment by teachers and school or college students is unlikely to be possible inside a factory workplace but the principles can be taught by attempting to detect emissions to the outside atmosphere from, say, local factory chimneys. Similar activities such as discovering the concentration of dangerous metals, like lead, in local dust (e.g. in the school playground) could be a way of encouraging interest in chemistry!
2. Day-release students at technical colleges could be encouraged to investigate health and safety problems in their own workplaces. It is possible to find out what safety regulations apply to their own factories, offices, etc., and to investigate whether they are being kept to both by individual workers and by the management.
3. School trips to local factories with talks by factory safety officers and trade-union representatives.

* Handpump air samplers and detector tubes are supplied by Draeger Normalair Ltd. of Blyth, Northumberland NE24 4RH, and by D. A. Pitman Ltd. of Weybridge, Surrey.

As to the questions around which these activities could be organized, three seem to be of some interest and importance in gaining a broader understanding of the causes and methods of prevention of accidents at, and diseases from, work.

Firstly, to what extent are *individual* workers to blame for accidents at work? Is it the case that as the Royal Society for the Prevention of Accidents (RoSPA) puts it, 'Everything possible is done to enable you to work in safety but your cooperation is essential' and that, therefore, accidents are due to such things as, 'horseplay and tomfoolery, carelessness and thoughtlessness, lack of concentration, lack of respect for oneself and others'? Or is it that, as Patrick Kinnersly puts it, 'Accidents are built into the workplace like holes in a pin table. Most accident prevention work starts when it is already too late. Exhortations and posters start to fly *after* the ill-conceived work system has been set up and accepted as normal'.

What limited evidence there is suggests that at least half the accidents which were reported to the relevant inspectors (the Factory Inspectorate) could have been prevented by reasonably practicable precautions. Of these, 'blame' for not taking the necessary and available precautions could be attached to management in 47 cent of the cases, to workpeople in 43 per cent of the cases and to both in 10 per cent of the cases.

Secondly, why is it that managements do not take these available precautions and therefore create situations in which accidents can happen? The recent report of the court of enquiry into the explosion at the chemical plant at Flixborough in which 28 workers were killed, commented that, 'If production is below target and profit is below budget, there are inevitably conflicting, albeit perhaps, unappreciated priorities'. So, are the goals of managers in factories (to keep production going, to maximise profits) such that safety regulations will inevitably be broken and accident situations produced? If machines, buildings and work arrangements have to be designed to the cheapest specification that will produce the most goods at the greatest profits, will safety, health and comfort be treated in effect, though not necessarily consciously, as bolt-on goodies?

Thirdly, are work hazard problems merely examples of the general problems associated with technological 'progress'? Are the hazards of working with asbestos and some asbestos products or were the deaths due to the explosion of the Nypro plant at Flixborough, the unfortunate but inevitable risks which must accompany any technological development? Certainly, such reasoning maintains, we must weigh the *dangers* to the workers involved in producing materials like, say, PVC or nylon against the *benefits* that the wide availability of such cheap plastics brings. But is this only half the story? A cheap material (like a plastic) will tend to replace a dear material (like a metal) if it can do the same job. This economic incentive can, therefore, increase the demand for a material, whose production might present some considerable hazard, by expanding its range of uses, perhaps unnecessarily. For example, asbestos, an extremely dangerous substance whose use and processing has theoretically been regulated in Britain since 1931, is used in a very large number of situations (from brake linings to pipe laggings) where other less hazardous but more expensive materials *could* be substituted. So in such cases it is the pressure on the individual company to expand the sale and range of its products, even if this would be undesirable on the basis of broader social criteria, which provides the background to the work hazard. The risk to the workers involved results from economic and social factors rather than from 'technology' alone.

It has only been possible to present the barest outline of the problems concerning safety at work and the possible response of the schools in 'preparing' their students for them. Further details and thought-provoking comments on the nature of current and potential work hazards can be found in the following three publications:

The Report of the Committee on Safety and Health at Work (The 'Robens Report'), Cmnd. 5034, HMSO, 1972;

Patrick Kinnersly, *The Hazards of Work: How to Fight Them*, Pluto Press, 1974;

Jeanne Stellman and Susan Daum, *Work is Dangerous to Your Health*, Vintage Books, 1973.

ENVIRONMENTAL POLLUTION

SIR FREDERICK WARNER

Partner of Cremer and Warner, Consultant Engineers

The interest in environmental problems suffers from fashion. It waxes and wanes with publicity, particularly of dramatic or horrifying experiences. On the popular side, the starting-point can almost be fixed with certainty as the publication in 1962 of *Silent Spring* by Rachel Carson. This was an emotional attack on the use of insecticides and was received coldly by the scientific establishment. The emotion was, however, a stimulus to an awakening public interest and was healthy in administering a shock to conventional anthropocentric attitudes. It helped to bring back the feelings that the whole of creation is connected and the disappearance of species is an impoverishment of us all. It focused attention on ecology as a study not just of interest to biologists but a human, an aesthetic and intellectual challenge.

This feeling needs to be retained as the reaction is now setting in that the major pollution problem is the state of life in which the majority of the world's population lives. This feeling came out strongly from the representatives of the Third World at the 1972 United National Conference on the Human Environment. It could be regarded as the high-point of the environmentalist movement with its basic criticism of industry and technology as the sources of all problems.

If the emphasis in environmental problems now shifts to what are social and economic phenomena, the ordinary problems of pollution will not go away. They remain in two categories: the short term and local or the long term and global. The short-term problems are those of the industrialized countries and are concerned with preserving or restoring the purity of rivers and oceans, reducing air pollution and noises and in clearing up dereliction. The long-term problems are the preservation of habitats and species, the safe storage of radioactive materials from nuclear reactors, possible climatic changes from atmospheric or stratospheric change and increasing local catastrophes from overcrowding of people and industrial activities.

In the short term, spectacular improvements have been made. The old type of London fog which caused the death of 4000 people in 1952 is now only a memory. The smoke and sulphur dioxide which caused it have been more than halved and every year the amount of winter sunshine over London increases. The Norwegians say that our tall power station chimneys take sulphur dioxide away from us and dump it on them as acid rain. This is not proved nor disproved although meteorologists have shown that the atmosphere above England is blown over Norway about a quarter of the year. The chemical reactions which cause the disappearance of trace gases are still not fully understood. Sulphur dioxide lives only about 24 hours in the air and is accompanied by ammonia in far greater quantities than required for neutralization.

There are also worries about lead from motor exhausts in city centres because of the effect on children. Epidemiologists have a difficult job to get statistics which are meaningful. The whole of pollution measurement ends up with statistical assessment because of the different dilutions which are reflected in the air or water receiving the pollutant.

The next 25 years will be exciting because of the changes in availability of the materials which cause pollution today. As petroleum production begins to fall off, there will be less private motoring, fewer oil-fired power stations. There may begin to be a drop in atmospheric carbon dioxide instead of an increase, so that climatologists will stop worrying about increased infra-red radiation. Fluorcarbons for aerosols will probably have disappeared whether or not their effect on stratospheric ozone (responsible for absorbing ultra-violet radiation) is significant compared with that of natural organic chlorides.

In water pollution, the increasing treatment of sewage should give purer rivers, and lakes not so liable to excessive growth and eutrophication. The Rhine as a river will be a great deal purer as industry cleans up its effluent and the French run out of potash in Alsace and no longer dump over 2 million tonnes a year of salt in the Rhine. The worry about mercury will have been forgotten as the only source will be from natural weathering of rocks or burning of coal. The Japanese horror at Minimato was exceptional and due to discharging catalyst from an obsolete process into a bay with little renewal of its water. Other industrial uses are being cleaned up or cut out.

Our own Thames estuary has come in 20 years from a foul, smelly state to an almost respectable state in spite of the prolonged drought in

1975. It has always had oxygen in it since the new sewage works at Barking were brought into operation. It will be interesting to see if the huge population of duck and waders which has grown by feeding on the rich mud will disappear as the mud becomes starved of the food for worms. It is unlikely to produce many salmon but by 2000 the Tyne should, unless salmon are wiped out by fishing off Greenland.

The significant areas of concern for pollution in the years ahead will be noise, litter, visual pollution from bad buildings or old tips and the odd accident when something escapes which should not. Some drinking water sources will have dissolved nitrates in amounts which would make the water unsuitable for babies and there will be scares about possible formation of nitrosamines, known to be carcinogenic. The worst air pollution will probably still be smoking but surely with some changes so as to reduce the current 50,000 deaths a year in the United Kingdom.

The attraction of environmental studies will remain because they involve ethics, aesthetics and intellect in the greater understanding of creation as a whole.

Transport

OVERLAND

BARON DUCKHAM

Senior Lecturer in Economic History in the University of Strathclyde

Who has not rejoiced in speed on the freeway or deplored scarred post-arterial surgery; admired the crisp functionalism of concrete flyovers or lamented the motiveless carnage of motorway madness? Have we not developed a schizophrenia through the clash of our belated 'environmentalist conscience' with our near veneration of quick individual transport and high pay-loads? Divested of moral cant, has not much of our social protest amounted in practice to 'new motorways by all means, but not past *my* house', a 'third London airport of course, but not on *this* site?', 'the automobile is the Frankenstein's monster of today's society—but leave *my* car alone'? Are not all of us large-scale consumers of transport? As the liberating dream of individual transport for everyman sours, man (it might seem) has little to regain but his legs. Is he to suffocate beneath a megapolitan mountain of cars or to strangle himself in a network of last-ditch bureaucratic restrictions?

Lamentation alone is sterile. Our adversaries are not the products of transport technology, but inertia, ignorance, defeatism; our allies are resolve, genuine involvement, imagination, richer assessment techniques to evaluate the consequences of our transport choices, and flexible attitudes to what must necessarily be constantly (if subtly) changing criteria for our decisions. We need to remember, too, that no solutions are final and that the challenges of transport technology, like any other form of social change, are continuous.

In approaching the question of planning for a just use of transport innovation, teachers can nurture a constructive, informed stance in their pupils in three ways:

1. By examining critically the administrative machinery of decision-making and implementation in overland transport, e.g. how are rail closures effected?, how are motorway routes planned?, how is change in urban transport shaped? Case studies—often with at least a modicum of published documentation—are invaluable. So too are opportunities for public debate; reading the briefs of pressure groups both for change and the status quo, enables young people to appreciate the planner's frustration with the slowness of procedure, as well as the objector's cynicism about the value of public inquiries. Gaming or simulated activities by pupils can provide a realistic analogue of stages in decision-making.

2. By enabling pupils to achieve some understanding of the basis for transport decisions. Can pupils be brought to appreciate the limitations of facts; are some more equal than others? Can they be led to realise the fallibility of predictive methods? (teachers might ponder the criticisms in *Changing Directions* of, say, the Transport and Road Research Laboratory's method). Can they come to understand the uses and limitations of cost–benefit analysis?

Careful map work and fieldwork ought to make possible a variety of simulation exercises covering some of the consequences of building a hypothetical motorway, by-pass, international airport, closing a railway or even, within stricter limits, the results of adopting alternative fuels/motive powers. Both class groups and 'loners' can usefully play the roles of various interests for and against a particular scheme, or of expert witnesses. Such role-playing ought to be as realistic as possible with, for example, children knowing something of the educational and career profile of the planner or witness they are representing. Realism of this nature can help bring to light the prejudices of each side.

Simulation makes the participants aware of the wide variety of criteria requiring consideration and of the difficult problems in assigning priorities where social, economic, technological, political, moral and aesthetic arguments jostle together. How do you measure the benefits of fuel savings or fewer man-hours, of speed, smoothness of traffic flow against (say) loss of agricultural acres and other amenities or the creation of noise and exhaust-fume pollutions? What are the merits of the government's rejection of the Roskill Report's recommendations, of the wild-life lobby trying to preserve Maplin or, for that matter, the well-organised middle-class case which saved Stansted?

Ideally both simulation and case studies (including children interviewing local people already affected by the march of transport technology) will teach pupils four things. First, the vast potential for either enrichment or impoverishment of society which transport change can bring; second, that well-conceived applications of transport technology can be assured only by a wholehearted determination to learn about the skills needed in exercising judicious choice; third, the very basic need to test the credentials of every pressure group or witness and question all data; and finally that there is never a perfect solution, only a compromise among priorities—a fact which underlines the requirement for scrupulous consideration of every claim.

Regrettably, both teachers and pupils exposed to much of the technical literature of modern transportation will also discover—painfully—that a fair number of planners and technologists appear to have lost (if they ever acquired) an ability to communicate simply and effectively. This is an important issue. Distinguished transport planners have demonstrated that complex questions *can* be made intelligible to the layman. We need to remember that unpalatable or ill thought-out designs can easily be disguised through the worst excesses of 'technicalese'.

3. By keeping a firm eye on 'tomorrow's world' in the realm of overland transport. Technological advance can quickly solve some of our present problems (e.g. some forms of pollution) and/or present us with new ones. Nowadays it is usually easier to change our potential technology than our attitudes (prejudices?)! There is no space here to enlarge on this all too obvious point. Suffice it to say that alternative sources of power for land transport, new concepts of 'super trains', possibilities of vertical lift-off for passenger aircraft and so on, will all demand new insights on our part. Already containerisation is producing a very real transport revolution for certain classes of freight on rail, road and sea (of which unfortunately, the juggernaut issue or dock labour troubles are often all the average pupil has heard of it).

In the past our decision-making in transport policy has taken too little account of probable technological innovation. All too often official planning has been a latecomer on the scene, trying desperately to cope with an existing technology rather than look ahead to possible future ones. Simulation and re-runs of case studies in the classroom can perhaps give a freer rein to the technological 'ifs' of tomorrow. With encouragement pupils may well show a degree of imaginative flair that often appears to have been denied to some of our planners.

Of course it needs emphasising that the kind of assignments suggested here cannot guarantee that the coming generation will acquire sufficient overall sensitivity to make fruitful decisions. One *can* guarantee that without such project work they will not.

Bibliography

The literature of modern transportation is vast and only a brief selection can be noted here. Teachers should be aware of some of the periodicals, though contributions are often very technical in presentation. See especially *Traffic Quarterly: an independent journal for better traffic* (Columbia University Press, New York); *Transportation:* an international journal devoted to the improvement of transport planning and practice, (Elsevier Scientific Publishing Company, Amsterdam); *Transportation Research: an International Journal* (Pergamon Press, Oxford); *Transportation Science* (Operations Research Society of America, Baltimore). For Great Britain premier mention must be made of the publications of the Transport and Road Research Laboratory, Crowthorne, Berkshire. A large and growing number of valuable reports exists; see especially: LR352 *Air Pollution from Road Traffic—a review of the present situation* (1970); LR357 *A Review of Road Traffic Noise* (1970); LR396 *Current Costs of Road Accidents in Great Britain* (1971); LR429 *Trends in the Transport of Freight in Great Britain* (1972); LR441 *Roads and the Environment* (1972); LR481 *Towards Safer Road Vehicles* (1972); LR543 *Forecasts of Traffic and Vehicles in Great Britain* (1973); LR566 *A Comparative Network Simulation of Different Methods of Traffic Restraint* (1973); 642 *Predicting Road Traffic Noise in the Rural Environment: a study of the A66 road improvement scheme in the Lake District* (1974); 650 *Forecasts of Vehicles and Traffic in Great Britain: 1974 revision* (1974); 659 *Environmental Simulator: progress report* (1974). Teachers are reminded that both the Centre for Environmental Studies, London, and the Centre for Transport Studies, Cranfield, Bedford, publish working and research papers on modern transport trends and problems.

Other reading

Aldcroft, D. H. *British Transport since 1914: an economic history*, Newton Abbot, 1975.

Antoniou, J. *Environmental Management: Planning for Traffic*, London and New York, 1971.

The British Railway Board's Review of Railway Policy, British Railways Board, 1973.

van den Burg, G., *Containerisation: a Modern Transport System*, 1969.

Changing Directions: The Report of the Independent Commission on Transport, 1974.

Drake, J. W., *The Administration of Transportation Modeling Projects*, Lexington, Toronto and London, 1973.

Energy for the Future, Institute of Fuel, 1973.

Hall, P., *London* 2000, 1963.

Highway Statistics 1973, HMSO, 1974.

Johnson, K. M. and Garnett, H. C., *The Economics of Containerisation*, 1971.
Nwaneri, V. C., Equity in cost–benefit analysis: a case study of the Third London Airport. *Journal of Transport Economics and Policy*, **4** (1970).
Personal Mobility and Transport, PEP broadsheet, 1973.
Physical Requirements of Transport Systems for Large Freight Containers, United Nations Dept. of Economic and Social Affairs, New York, 1973.
Roads of the Future: the new inter-urban plan for England, Cmnd. 4369, HMSO, 1970.
Roads in England 1973–4, HMSO, 1974.
Scottish Roads in the 1970s, Cmnd. 3953, HMSO, 1969.
Sharp, C., *Living with the Lorry*, 1973.
Sharp, C., *Transport Economics*, 1973.
Traffic in Towns: Reports of the Working and Steering Groups, HMSO, 1963.
Transport in the year AD 2000, Institute of Mechanical Engineers, 1965.
Urban Transport Planning, House of Commons Paper 57, Session 1972–3, 1973.

ON WATER

FRANCIS EVANS

Department of History and Communications Studies, Sheffield Polytechnic

Rarely do we ask the question 'Why do it this way?'; 'Why use ships?'; 'Why transport goods overland on canals, when this involves the labour of transporting goods back onto land vehicles at the end of the journey?'; 'Why choose water?'

Seventeen-fifty marked the beginning of modern transport development in Britain. With a remarkably long, indented sea coast in proportion to her land area and a large number of very bad roads, it was unthinkable to carry bulk goods overland. These travelled by sea: coals were shipped from Newcastle to London, for instance. Wherever possible rivers were used and sometimes 'improved' to allow the passage of boats. Then, after 1768, canals were dug.

Why all this change? The reason lay in the power needed to drag loads on bad eighteenth-century roads; a horse could pull perhaps a ton in an unsprung cart and before Macadam the roads were too bad for springs to be used. On a canal, the same horse could pull 50 tons. On a bad road, the horse virtually had to lift the load over bumps: there are no bumps on water, and resistance is small at low speeds. Horses were expensive to keep, but wind power was free. By the eighteenth century, sailing ships were highly developed. They could sail very close to the wind, easily within 45°; they had effective simple sail systems which reduced the manpower needed to control a vessel.

Nowadays we tend to ignore windpower because it requires bulky equipment to capture it, and because it is unpredictable. We think sail-power is old fashioned. Yet *Cutty Sark* could sail at $17\frac{1}{2}$ knots. On one occasion she logged 2164 miles in 6 days: on another, 3457 miles in 11 days. Accurate models of ships are sometimes tested in big water tanks to measure the behaviour of their hull shape; testing a hull model of *Cutty Sark* showed that at $16\frac{1}{2}$ knots she would have needed engines giving 4000 indicated horsepower—but all this came free with the wind.

Steam power killed the canals and the clippers. A horse could pull 50 tons on a canal, 1 ton on a road, and 10 tons on rails but only at $2\frac{1}{2}$ mph. With higher speeds water resistance rose sharply; at only 5 mph a canal barge needed the same tractive effort per ton as a railway load. By 1829 steam power could work at 20 mph or more; at sea it gave reliable service which wind could not equal.

Canal transport had other insuperable drawbacks at that time. Locks offered the biggest problem. Despite great efforts to follow natural contours, canals had to change level and every 12 feet of rise or fall necessited a lock. It took about 10 minutes to pass through a lock and some 200 tons of water were displaced in the process. Unlike natural waterways, canals need to be supplied with water, and in summer this was often scarce at a time when water-powered industries were also demanding it. Desperate efforts were made to find substitutes—lifts, steep sections of railways, and even cylindrical barge carriers capable of sinking and surfacing like submarines. But early nineteenth century technology simply could not solve the problem, and the railways won.

Today, we should think again, for the world is fast changing. Coal and oil are not the cheap commodities they were when sail and canals were ousted: the fuel costs of speed are no longer quite so acceptable. It would be interesting therefore to consider what modern technology could offer to a modern sailing cargo ship. Since the days of *Cutty Sark*, a whole science of aerodynamics has developed—and what is a sail but a kind of aerofoil set on end? There would be no need of men to climb masts or haul on ropes if advances in cybernetics, control and servo-devices were applied to windpower. Even the idea of the ship as a box to put other boxes in, could be rethought. Why should not boxes that float be used, even if they finish their journey by train or lorry? Or again, why use boxes for liquids when the sea would treat plastic tubes or bags gently, and these call for no heavy rigid structure?

A powerful feature of modern technology is the way it has freed us from tradition: instead of making small improvements to the usual form of a machine, we have acquired the art of analysing the function we wish to accomplish, and then radically designing the means to do so. In the past this sort of rethinking often took a long time: Richard Trevithick's first railway engine, in 1805, was 'led' by its driver who went on foot beside it, opening and closing the steam valve as if he were leading a horse; the first plastics were made as artificial ivory or wood; early electric motors had beams and cranks and flywheels in 1840, just like James Watt's steam engine. Nowadays we rethink our designs much more radically. For instance, a normal large ship loses 80 per cent of its power at 30 to 50 mph in dragging its sides through the water. So Professor Thring suggests a ship with moving sides made from two large rubber belts moving rather like horizontal tank tracks. This would eliminate a good deal of friction: it could also do away with propellers for it would be possible to steer the ship by running the tracks at different speeds. Or again, large transport submarines could be propelled more economically than conventional ships; large hydrofoils would also expend less power on forming waves, but these present too many problems.

New technologies, however, are not easily accepted. It is not only the Technical Whigs or individuals with rigid ideas who prevent improvement. An adequate technology does not just mean the ability to make things; it also implies the capacity to accept new ideas, to realise their implications and to make wise decisions. There are many examples of new inventions which were rejected in the land of their origin but taken up elsewhere. France and Holland made many of the advances in hydraulics and canals which chiefly helped the dynamic England of the eighteenth century. French ship design was better than British in the Napoleonic wars, and captured French vessels were often copied exactly in English shipyards. Yet there is no doubt that England's use of the ships was altogether more forceful and decisive.

All this is another way of saying that technology is meaningless when abstracted from the context of humanity. In many respects society determines the form and functions of technology and is, in turn, altered by it. We can do almost anything, but our time is growing shorter: this is the paradox which forces us to consider very carefully what we do.

Bibliography

H. Butterfield, *The Whig Interpretation of History.*
C. Nepean Longridge, *The Anatomy of Nelson's Ships.*
H. I. Chapelle, *The Search for Speed under Sail.*
L. T. C. Rolt, *Navigable Waterways.*
C. Hadfield, *British Canals.*
B. Landstrom, *The Ship.*
M. Lewis, *The Navy of Britain.*
M. W. Thring, *Man, Machines and Tomorrow.*
Sir George Thomson, *The Foreseeable Future.*

Suggestion for practical work

Use simple equipment and some ingenuity to measure the traction required to move weights on rails (real or model) and various surfaces, including water. What shapes pass most easily through the water?

Communications

THE MASS MEDIA

GEOFFREY SIMS

Vice-Chancellor, formerly Professor of Electronics, University of Sheffield

With the invention of the telephone one man was able to talk to another —even though he was separated from his fellow by a great distance. With the discovery that radio waves could be used to carry messages without the need for wires, the idea of 'Broadcasting' became a reality. One man could now communicate with great numbers of people at once, provided that they all possessed 'wireless' sets.

Thus 'information', 'education' and 'entertainment' could be sent 'over the air' to countries in all parts of the world. Radio and the telephone had effectively, therefore, made the world a much 'smaller' place: no longer did you have to take your message, or even post it, to the recipient; no longer did you have to 'go out' to hear music or a talk—you could 'receive', almost instantaneously, in your own home. People's habits were already beginning to change.

Then came television with the ability to broadcast, sending either over the air or by cable, *visual* information. The picture, as well as the sound, was available in your home. Habits changed even more; why go out to a football match when you could see it in your home; why go out to a theatre or to a concert? Was this an encouragement for people to become lazy, were they missing something, or did it mean that they just had more time to spare in which they could do other things?

At first television was only available to the few, for sets were expensive —but more importantly the 'waves' that were used to carry television pictures could only travel in straight lines and, hence, if there was a mountain, or other obstacle, between your home and the transmitting station, the signal could not travel round the mountain as would a 'radio' signal. To make television available to the many it was necessary to receive the signal by a special aerial on top of the obstacle and retransmit it either, through the air, or by cable, to the home. To establish a Nationwide television service thus took time and was, furthermore, an expensive process. Even today not everyone in Britain can receive a television programme and many do not receive all of the three 'channels' at present in use.

The last big step forward was when it became possible to send pictures direct from one country to another, by transmitting them up to a satellite which was either hovering over the earth or orbiting around it, and using it to retransmit the programme back to earth again, in a way which made it available over vast areas. Now we could not only hear about what was occurring in other countries but we could see it too, *as it happened*. The world thus became smaller still.

We *learned* through television, and the ability which it gave us, to *see* how others lived, and it brought to our sight much of what had been previously available only in books about geography, and the printed word had thus become 'alive'. World happenings of great significance were, perhaps, rather taken for granted as we watched—or were they? Were we really less curious to see for ourselves, to read, or to investigate or had television, on the other hand, actually made us not only more knowledgeable, but more curious to find out about things—by whatever means?

These are some of the questions that we ask ourselves—but there are others of even wider significance which we shall consider when we have looked into the future a little further.

At the moment in Britain only three television programmes are broadcast though we can, even now, foresee a future in which perhaps as many as six could be transmitted directly over the air and perhaps four more could be received indirectly via satellites—possibly from other countries. What should we do with so many channels available to us? Could we fill them all with entertainment, more news, more education? Should we make more 'television time' available for the public to use themselves, for discussions or broadcasting their own programmes—is public participation important?

In some countries, e.g. Canada and Holland, many homes are connected to receive television by cable and a modern cable makes it possible for each home to receive perhaps a dozen or more programmes. Some of these systems allow the listener to communicate back to the studio, e.g. to vote on programmes, or to answer questions. What possibilities does this raise? Is such a facility desirable or necessary? Such a system would be expensive to install in Britain, but could be attained—if we could afford it! Making any kind of programme costs money and needs talented people to perform and produce it. Does this impose a limitation on how much useful mass communication we can usefully have also? How important is it to devote more of our national resources to more television, or is it of greater importance to have more roads, more schools, more . . . ?

These are difficult questions particularly when we can foresee within the next 20 years the possibility of utilising new kinds of glassfibre cable which could give us the capacity to broadcast an almost unlimited number of television transmissions as well as carrying all of the telephone and data services we would need. It would be possible to use such a cable to obtain on your screen, merely by pressing a button, information from the local library, the latest football results, or you could 'dial' the 'computer bank' so that a particular film you wanted to see, was transmitted. Would this be a good thing? Is it worth spending our limited national capital on? Would it change the pattern of employment, e.g. in libraries, the film industry, etc.? Further, would it be right that someone could buy up the right to broadcast, say, the Cup Final and transmit it only to a select few who had agreed to pay extra for it?

This latter question is a big one, for at the present time the Government does not allow 'anyone' to transmit programmes, whether by cable or over the air, as it has wished to prevent the public being denied access to things which are clearly of interest to all. Should the Government keep such control, should it be in the hands of some other body, or, having installed the capacity to transmit, should it then be a system free for all to use?

These are some of the questions with which the future technological possibilities challenge us. All of these things *could* be done, but *should* they be done and how great a priority should we give them in terms of our national resources?

This question has deliberately been put in this manner because, in broadcasting, there are almost no bounds to what is *technologically* possible in the future. We have so far been slow to appreciate the differences that 'instant communications' have made to our lives. Furthermore, the responsibility which rests on the shoulders of the broadcaster, to ensure that whatever he broadcasts is absolutely accurate does not represent a biased point of view, or does not mislead, is enormous. We rely tremendously on 'the media' for information and what it features we tend to regard as both true, representative and important. Can the broadcasters really judge what is important; can they provide a 'neutral' view; is it right that they should try to be neutral?

Broadcasters are continually being criticised for transmitting too many programmes concerned with sex, violence, or even advertisements which are said to adversely influence the public. They are accused of presenting a distorted view of youth as being, at best, amoral and many young people would protest strongly, saying 'we are not like this at all!'

Perhaps the greatest challenge facing the broadcaster, then, is that of how far he *can* be neutral. Should he not be trying to use the immense power of television or radio to lead people to a greater sense of moral purpose? If television can be thought to be an influence for evil, can it not be equally an influence for good, whether it is the good of individuals, the Nation or Mankind overall?

Who then should control the use of this powerful tool which can bring to us such diverse messages and should its use be neutral? If it is not to be 'neutral' how can we ensure that what it offers is going to be for 'good'?

In many respects this is one of the most important questions which we face for the future. Technology has offered us the possibility of unlimited communication: the original objectives of broadcasting were 'to inform, to educate and to entertain'. Are these objectives still appropriate and sufficient? In whose hands should the answer to this question lie and how should the future development of the mass media be financed if it is agreed that their influence is as important as it currently seems to be?

References

I. There are few, if any, books which refer to the whole range of possible discussion topics mentioned in this article. Perhaps the best sources of information as to what is happening currently are the annual publications of the BBC and the IBA under the titles

1. *BBC Handbook* (1975, 1976, etc.).

2. Independent Broadcasting Authority ITV (1975, 1976, etc.): *A Guide to Independent Television.*

II. The BBC and IBA publish many pamphlets and other literature concerned with all aspects of broadcasting, including planning, which they are usually happy to supply on request, e.g.

(a) Curran, Charles, *The BBC in the eighties: the relationship between broadcasting policy, programme needs and technological potential, BBC,* 1972.
(b) Redmond, James, *Broadcasting: the developing technology,* BBC, 1974.
(c) Weigall, Denny B., *Satellites—present use and future ideas in broadcasting,* BBC, 1968.
(d) BBC, The use of radio frequencies for sound and television broadcasting in the United Kingdom: a paper prepared for the BBC for its General Advisory Council, BBC, 1974.
(e) Pawley, E., *BBC Engineering 1922–1972,* BBC, 1972.

III, The following publications, mostly books, also discuss some relevant ideas:

(a) Halloran, James D., *the Effects of Television,* Panther, 1970.
(b) Smith, Anthony, *British Broadcasting,* David & Charles, 1974.
(c) Social Morality Council, *The Future of Broadcasting* (includes 4th channel and cablevision). Eyre Methuen, 1973.
(d) Williams, J. I. R. *Television: technology and cultural form,* Fontana, 1974.
(e) International Broadcasting Institute, *The new communications technology and its social implications: report of a symposium by R. P. Morgan,* The Institute, 1971.
(f) Institution of Electrical Engineers, *Broadcasting Technology—past, present and future.* A programme of lectures ..., IEE, 1973.

IV. A major work dealing with the development of broadcasting generally which provides a comprehensive background is

The History of Broadcasting by Asa Briggs.

Three volumes: 1. *The Birth of Broadcasting,* 1961.
2. *The Golden Age of Wireless,* 1965.
3. *The War of Words,* 1970.

Published by Oxford University Press.

DREAMS OF COMPUTERS

CHRISTOPHER EVANS

National Physical Laboratory

and

PETER DUTTON

Sheffield Region Centre for Science and Technology

Anyone who has worked amongst computers over the past decade and watched their evolution is almost bound to become a computer-fan, but being also a fan of human beings causes one to watch the further evolution of computers with the greatest possible interest in moods which swing between glorious optimism and a sense of distinct disquiet. The optimism and the disquiet are both caused by the awareness of the extremely rapid pace of computer development.

The main part of this article is about dreams of computers in the future but the title also accurately reflects the drawing of parallels between computers (which are complicated data-processing devices whose principles of operation are largely understood) and brains (which are even more complicated data-processing devices whose principles of operation are distressingly mysterious).

Whereas biological brains have evolved as the result of constant pressures from the environment over the course of an immense period of time, computers have evolved to their present level in not much more than a quarter of a century. Unlike brains, computers have evolved not because of the environment only but by the brain and hand of man. This has led to their phenomenally rapid rate of intellectual growth. Already in some, if not many, respects computers are man's intellectual equals. But in two aspects they are man's superiors! these are the aspects of speed and accuracy of calculation. Almost all computer scientists accept that the present rate of development is accelerating which will bring computers up to or in advance of man's level on more and more counts within a small number of years.

One of the pioneers in the existing area of study known as artificial intelligence is the distinguished mathematician, Professor I. J. Good of Virginia State University, U.S.A. He has introduced the concept of the Ultra-intelligent machine (UIM) which, by his definition, will perform *all* human intellectual functions as well as any human being. He predicts these may be produced in the 1990s and certainly by the end of the century. No doubt some experts would consider this prediction to be optimistic (is pessimistic the right word here?), but even if Professor Good is out by a decade or so, does it matter very much? The first UIMs will be in operation while most of the present school population are still alive.

Even when the UIMs do arrive, will they really make a difference and in any case will it really be worthwhile building them? Some years ago, Lord Bowden was asked why one should spend millions of pounds on creating a computer which was the intellectual equivalent of man when there were so many of these computers—admittedly in biological form—walking around on the face of the earth, and which were so much easier and so much more pleasant to manufacture. Leaving aside for a moment one obvious comment about the enormous difficulty that is experienced in programming these biological computers—it takes 20 years at least in most cases before they even come up to average—there is a more important point that has been missed. It is this.

Whereas there appears to be a practical and possibly theoretical upper limit to human intelligence using the present biological brain-computers, there appears to be no reason why there should be a practical or theoretical upper limit to artificial computer intelligence. The UIMs are not likely to rest content once they have reached an IQ of 120 or thereabouts. One of the first jobs that any self-respecting computer scientist will set the UIMs to do will be to raise their own IQs, since there is little possibility of their doing much to raise the IQs of human beings. Perhaps progress will be slow in the early years as they struggle up through the 130s and 140s, but of course the brighter they get the better one would expect them to be at raising their own intelligence. By the time the UIMs' IQs get to the 200s, then one might expect the graph to start going up exponentially. This is the cause for both optimism and for disquiet.

Will the UIMs actually be encouraged to devote their energies to improving themselves in this extraordinary way? It can be argued that man can decide how far to proceed and can stop further development when the IQs have reached some agreed limit. But what might that be? 90? 120? 150? 200? At what point will the obvious gains of having very bright computers available be outweighed by the somewhat less obvious losses? One guess is that man will be unable to resist the temptation to develop UIMs and their successors. The reason for this guess is a very practical one.

By modifying his environment, developing his technology and multiplying his social relationships, man is swiftly reaching the stage where his universe will become too complicated for his central nervous system. It was once a dream of science that the more that was discovered about the world the simpler it would appear to be. The reverse has turned out to be devastatingly true—the more that is learnt about the universe, the more peculiar and complicated it becomes and so the chance for manipulating it, controlling it and understanding it becomes less. Man's very survival as a species will depend on an extremely rapid expansion in computer technology employing the potentially unlimited intellectual power of computers to assist him into the increasingly elaborate future that he is preparing for himself.

One form of the disquiet felt by the possible exponential growth of the intelligence of computers is summarised in a marvellous, very short science fiction story by Frederick Brown called 'Answer'. The time is in the far distant future when the heads of numerous intergalactic federations meet for the inaugural linking together of all the computers on a hundred thousand planets. There is intergalactic TV coverage of course. The President of the Universe solemnly throws the switch linking the systems together. He then steps forward to ask the ceremonial first question, previously selected for him by a team of the galaxies' greatest philosophers and scientists. 'Is there a God?' he asks. There is a moment's pause until a great voice booms forth. 'There is NOW.' The President pales and then leaps forward to switch the computer off. Before he can do so, a bolt of lightning strikes him dead.

Despite the chill that this story produces, the results of the steady moves towards UIMs will probably not lead to such an apocalyptic event but will more likely be more subtle. There may come a time when computers tell man something about the universe that it will not please him to know. (What that might be could be a source of much speculation.) One thing that is more certain is that the UIMs will eventually decide to convey their intelligence to man, i.e. there will be a true man–computer conversation. For the first time therefore man will find himself in contact with another being, non-biological in origin, with which he can communicate, exchange ideas and which will be at the very least his intellectual equal. Man has always been proud of his intelligence and the creative capacity that goes with it—abilities that he believes make him truly unique. The real question in the dream about computers is therefore not any development that may change man's life style, environment or general fund of knowledge but rather how will computers change man's image of himself?

If this analysis of future trends in the growth of computers is correct, then Education has an enormous task ahead to prepare, inform and involve people with those working with computers. This task will stretch from school to higher and further education and into many forms of adult education. It will be a continuous task that will need a very high order of imaginative thinking by teachers if it is to have any chance of success. The sheer size of the problem as well as the innate importance for all mankind requires many, many teachers to be involved, not just those who are using the machines directly in their work.

Accepting the reasons why teachers should become involved, it will be impracticable, nay probably impossible, to study the details of the growth of computers, certainly with school pupils. The approach most likely to lead to a successful understanding of the problem of the use of these machines will be to involve the pupils and students with a consideration of the *implications of the uses* rather than the details of either the machines or the users themselves. For example, it is quite feasible to involve young people in using their imagination to forecast the growth of the use of credit cards leading to the possible eventual introduction of a cash-less society. Such uses would *have* to be computer-controlled because of the size of the problem and the speed with which responses would be required. This raises questions about the control of the controlling computers which are readily understood by school pupils, yet the problem is not so far-fetched that they fail to realise the present, let alone future, implications. It is suggested that such involvement is practical, possible and very necessary for all school pupils if the interactions in the future between people and UIMs are to stand a chance of being peaceful.

The Future of Man

SOME THOUGHTS ON SOME ASPECTS OF BIO-ENGINEERING

GEORGE B. BALLESTER

Woodrow Wilson Fellow, Linacre College, Oxford University

'Bio-Engineering' is the engineering (in the senses of mechanical and chemical engineering) of living organisms (in the sense of biology). As engineering concerns dead materials, and biology living organisms, there is an apparent contradiction in the term. 'Bio-Engineering' is usually divided into four branches—genetic, neurophysiological, medical, and behavioural.

After Charles Darwin's *Origin of Species* (1859) the science of genetics gathered momentum; and arguments were made for genetic engineering (eugenics), especially by Francis Galton (two articles in *MacMillan's Magazine*, 1865; *Hereditary Genius*, 1869; *Inquiry into Human Faculty*, 1883). His defence of eugenics is summarized in *Memories of my Life*, 1908, p. 323:

> Man is gifted with pity and other kindly feelings; he has also the power of preventing many kinds of suffering. I conceive it to fall well within his province to replace Natural Selection by other processes that are more merciful and not less effective. This is precisely the aim of Eugenics. Its first object is to check the birth-rate of the unfit, instead of allowing them to come into being, though doomed in large numbers to perish prematurely. The second object is the improvement of the race by furthering the productivity of the Fit by early marriages and the healthful rearing of children. Natural Selection rests upon excessive production and wholesale destruction; Eugenics brings no more individuals into the world than can be properly cared for, and those only of the best stock.

Under Galton's influence a research fellowship (1905), a research scholarship (1907), and a laboratory (1909) in eugenics were founded at the University of London. Karl Pearson, Galton's biographer, continued the argument with *The Scope and Importance to the State of a Science of National Eugenics* (1911). The movement, unfortunately, took an ugly turn with Chamberlain's *Foundations of the Nineteenth Century* (1911), Gobineau's *The Inequality of the Human Races* (1915), and Nietzsche's philosophy of the *Übermensch*, flowering in Nazi Germany's genocide of the Jewish peoples. In recent years, the de-coding of DNA and RNA by Watson and Crick (1962) gave a biochemical basis to eugenics. The primary dilemma has always been how to objectively define, aside from the carrying of hereditary diseases, the 'Unfit'. Many great men have not been fit in the Aryan sense; and benevolence is not necessarily correlated with intelligence quotient. There is a real danger of unscrupulous politicians eugenically exterminating their opponents.

In the late nineteenth century, neurophysiologists astounded the public with feats of neurophysiological engineering. In 1881 Goltz exhibited dogs—unable to hunger, lust, or remember—because their cerebrums had been removed. At the turn of the century Sir Charles Sherrington found that monkeys with transected spinal cords lost all reflexes for several hours; and that animals, whose cerebral hemispheres had been removed, moved their diagonal limbs in pairs, harmoniously in the same direction. In 1927 Pavlov demonstrated his ability to control the behaviour of hungry dogs by establishing conditioned reflexes. Such feats raise questions of the mind's connection with the body and of the reality of free will. There is a real danger of unscrupulous politicians using neurophysiological engineering to control thought and behaviour.

With the introduction of anaesthesia and antiseptics in the second half of the nineteenth century, there was a large increase in the number of surgical operations, resulting in technical improvements. Ironically, in our century, the horrible casualties of modern warfare have led to breakthroughs in the engineering of artificial limbs, tissues and organs. Recently surgeons have been able to transplant organs from one body to another. Transplants raise questions of when a person is actually dead, the ethics of obtaining donors, and of personal identity. It is interesting to note that in Mary Shelley's *Frankenstein* (1818) the monster was the hero.

Behavioural engineering, a twentieth-century phenomenon, was built upon the associationist psychology, reflexology, and animal psychology of the previous century. Behaviourism, as a self-conscious propaganda movement, is usually dated from J. B. Watson's 'Psychology as the Behaviorist Views It', 1913: 'It is possible to write a psychology, to define it as Pillsbury does (as the 'science of behavior'), and never go back upon the definition: never to use the terms consciousness, mental states, mind, context, will, imagery, and the like . . .' The new psychology would be based on observed stimulus-response bonds, the engineering, on their manipulation. A modern behaviourist, with political influence, B. F. Skinner argues for State behavioural engineering in *Beyond Freedom and Dignity*, 1971: 'What is needed is more control, not less and this is itself an engineering problem of the first importance' (p. 173). Another area of behavioural engineering is in ethology—notably Lorenz, Ardrey and Tinbergen—where instinctive behaviour is rechannelled. Behavioural engineering raises questions of the mind–body relation, free will, and of the Darwin–Wallace debate about whether the higher mental faculties of man can be wholly accounted for biologically. There is a real danger of unscrupulous politicians using behavioural engineering to establish a totalitarian society.

Bio-engineering raises political issues. If one respects the dignity and freedom (however limited) of each man, one could not tolerate State Bio-engineering. History exhibits a variety of motives for State Bio-engineers—e.g. the institutionalization of Utopia, the alleviation of unnecessary suffering, conceit, the creation of the 'new man', the genocide of the group of people one abhors, the end of social misfits, the control of one's political opponents, etc.

Bio-engineering raises philosophical problems. Polanyi has argued that 'Bio-engineering' is a self-contradiction; hence, impossible. Although physical and chemical analyses are necessary for competent biology, they are not sufficient because they fail to capture what separates living organisms from dead materials. Such analyses are not even sufficient for understanding machines, i.e. a space traveller or an archaeologist discovering a hitherto unknown machine would not understand it after such analyses, because he would not know its purpose. Bunge has denied the force of such arguments and has argued that biology is not methodologically unique. Waismann has shown that there may be good reason to doubt whether science proves the existence of a determinist Nature sufficient for the ultimate success of Bio-engineering.

The determinism of classical mechanics is a utopian idealization based upon presuming unlimited measurement accuracy—hardly a scientific presumption, if measuring instruments have an atomic structure. The Copenhagen interpretation of Quantum Mechanics denies strict determinism. General Relativity need not be interpreted in Weyl's determinist fashion.

Bibliography

1. *General Background*

Nordenskiöld, Erik, *The History of Biology—a Survey*, trans. L. B. Eyre, Knopf, N.Y. 1928.

Derry, T. K. and Williams, T. I., *A Short History of Technology*, Oxford U.P., 1960.

Singer, Charles and Underwood, C. A., *A Short History of Medicine*, 2nd ed., Clarendon Press, Oxford, 1962.

Boring, Edwin, *A History of Experimental Psychology*, 2nd ed., Appleton, N.Y., 1950.

2. *Genetic Engineering*

Darlington, C. D., *Genetics and Man*, Allen & Unwin, London, 1964.

Galton, Francis, 'Chapter XXI—Race Improvement', *Memories of my Life*, Methuen, London, 1908.

Pearson, Karl, *The Scope and Importance to the State of Science of National Eugenics*, Univ. of London, 1911.

Blacker, E. P., *Eugenics—Galton and After*, Duckworth, London, 1952.

Weinrich, Max, *Hitler's Professors*, Yiddish Scientific Institute, N.Y., 1946.

3. *Neurophysiological Engineering*

Walker, A. (editor), *A History of Neurological Surgery*. Bailliere, Tindex & Cox, London, 1951.

Pavlov, I., *Conditioned Reflexes* (1927), trans. G. V. Antruj, Dover, N.Y., 1960.

Sherrington, Sir Charles, *The Integrative Action of the Nervous System*, Scribners, N.Y., 1906.

Eccles, Sir John, *The Neurophysiological Basis of Mind*, Clarendon Press, Oxford, 1953.

Campbell, Keith, *Body and Mind*, Doubleday, N.Y., 1970.

4. *Medical Engineering*

Life, death and medicine issue, *Scient. Amer.* **299**, no. 3. (Sept. 1973).

Lenihan, John, *Human Engineering—the Body Re-examined*, Weidenfeld and Nicolson, London, 1974.

Kung, R. M. and Fehr, H. (editors), *The Challenge of Life—Biomedical Progress and Human Values*, Berhauser Verlag, Basel, 1972.

Miller, Henry, *Medicine and Society*, Oxford U.P., 1973.

5. *Behavioural Engineering*

Watson, J. B., Psychology as the behaviorist views it, *Psych. Rev.*, **20** (1913), 158–77

Skinner, B. F., *Science and Human Behavior*, Free Press, N.Y., 1953.

Eibl-Eibesfeldt, I., *Ethology—the Biology of Behavior*, trans. C. Klinghaner, Holt, Rineholt & Winston, N.Y., 1970.

Borger, R. and Cioffi, F., *Explanation in the Behavioural Sciences*, Camb. U.P.,1970.

6. *Bio-engineering and Politics*

Mill, J. S., On Liberty, *The Six Great Humanistic Essays of John Stuart Mill*, Washington Square Mess, N.Y., 1969.

Williams, Raymond, Social Darwinism, in *The Limits of Human Nature*, edit. J. Benthall, Allen Lane, 1973.

Hitler, Adolf, *Mein Kampf*, Reymond and Hutchinson, London, 1940.

Skinner, B. F., *Beyond Freedom and Dignity*, Jonathan Cape, London, 1972.

Commoner, Barry, *Science and Survival*, Viking Press, N.Y., 1967.

7. *Bio-engineering and Philosophy*

Johnstone, James, *The Philosophy of Biology*, Cambridge, U.P., 1914.

Polanyi, Michael, Science and Man, Nuffield Lecture, *Proc. Royal Soc. Med.* **63** (Sept. 1970), 969–976.

Bunge, Mario, 3. Is biology methodologically unique? *Method, Model and Matter*, D. Reidel, Dordecht and Boston, 1973.

Waismann, F., VI. The decline and fall of causality. *How I see Philosophy*, edit. R. Harré MacMillan, London, 1968.

Our Heritage

INDUSTRIAL ARCHAEOLOGY

BARON BUCHANAN

Director of Centre for the Study of the History of Technology, University of Bath

A remarkable phenomenon of the last few years has been the enthusiasm generated by such activities as the restoration of abandoned canals and the conversion of redundant railway track into busy lines for steam-driven trains. All over Britain groups of dedicated individuals have devoted large amounts of time and money to these and similar ventures, and there can be no doubt that their enterprise has given a tremendous boost to traditional museums as well as encouraging the development of new industrial museums and open-air museums. They have also added a new dimension to topographical interest and provided some thriving tourist attractions. The collective concern of these efforts has become the subject matter of 'Industrial Archaeology', a term which was virtually unknown before 1955 but which is now recognised in an extensive literature covering the main regions of Britain (there are 18 titles, including associated volumes, at present in the David & Charles series on 'The Industrial Archaeology of the British Isles'), the major industrial categories (there are 12 titles at present in the Longmans/Allen Lane series on Industrial Archaeology), and almost every canal and railway which has ever existed in Britain, as well as several general guides to the subject (K. Hudson: *Industrial Archaeology—An Introduction*, Baker 1963; A. Raistrick: *Industrial Archaeology—An Historical Survey*, Eyre, Methuen, 1972; R. A. Buchanan: *Industrial Archaeology in Britain*, Penguin, 1972; Neil Cossons: *The BP Book of Industrial Archaeology*, David & Charles, 1975).

Industrial Archaeology has thus certainly arrived on the library shelves. What is not so certain is the significance of its arrival. In an attempt to be systematic about its nature and its objectives, several of us have tried to frame a definition of the subject. My own attempt (see *Industrial Archaeology in Britain*, p. 20) emphasises the functions of investigating, recording, surveying and preserving industrial monuments. For purposes of this definition, an 'industrial monument' is any relic of an obsolete (or becoming obsolete) industry, transport system, or associated housing or other structures, of any age, although inevitably the greatest volume of such material comes from the recent past. The functions of investigating, recording, surveying and preserving proceed from the general to the particular: the initial exploration and preliminary recording may apply to all industrial monuments, the need for surveying is restricted to the more interesting monuments, and preservation is available in only a select few cases. It is the monuments considered for preservation, however, which receive most attention, either on account of the publicity gained for them (Brighton West Pier, St. Pancras Station, Conway Suspension Bridge, are prominent recent examples) by preservationist groups, or because of the attention given to them by the Department of the Environment or the Science Museum or other public bodies involved in administering protective legislation or funds for preservation. The development of a consistent policy sympathetic towards the preservation of industrial monuments has been an objective pursued by the Council for British Archaeology Industrial Archaeology Research Committee, and by the newly formed Association for Industrial Archaeology.

This bias towards preservation policy is natural enough at a time when obsolete industrial relics are under great pressure from programmes of urban renewal, highway construction, and the expansion of new industries, all of which have given urgency to the need to preserve as much as possible in order to retain a worth-while industrial heritage. It is also part of a widespread conservationist concern which is currently influencing many aspects of policy-making in the industrialized world, from urban planning to environmental ecology. In this respect, the rise of Industrial Archaeology can be identified with a general increase in historical consciousness and a concern to retain the best of the past. But many scholars, while recognising the practical services of Industrial

Archaeology in preserving important monuments and even in providing a valuable extension to leisure amenities, are still not satisfied with its academic performance. They want to know what does Industrial Archaeology prove?

The answer to this question depends upon the point of view from which it is asked. To the engineer, the study of an ancient machine can reveal how it works and what techniques were used in its construction: to the architect, the study of an industrial building can reveal much useful information which drawings alone can not easily convey: to the geographer, the study of changing industrial processes can help to elucidate settlement patterns and urban landscapes: to the historian, the study of the surviving artefacts of an industry can give depth to the understanding of that industry. In all these ways, Industrial Archaeology is something of a teaching aid: a short-cut to knowledge which could be achieved, albeit with greater difficulty, without it. It is helpful, but not essential. At best, it may be a positive aid to learning, and at worst it may be light relief, like some colourful slides at the end of a tedious lecture.

Yet in some ways Industrial Archaeology can do more than this, and shows real promise of growing into a distinct academic discipline, capable of advancing the frontiers of knowledge. Many industrial archaeologists, including myself, have not accepted the limitations of the definition as I have already given it, but have insisted on adding to it the importance of interpreting the evidence collected by industrial archaeological research in the context of economic, social and technological history. In other words, they have attempted to use the insights of Industrial Archaeology in order to modify the existing historical interpretations. It should be observed that Industrial Archaeology is part of a broader category of historical research which might well be called 'Physical History' because it relies on physical rather than on documentary evidence. In this it makes use of special techniques such as aerial photography, excavation and metallurgical analysis, which are shared with other aspects of Physical History like Archaeology and Historical Geography. It is this preoccupation with the physical evidence which justifies the word 'archaeology' in the term 'Industrial Archaeology', and by comparison with such adjacent studies it is possible to indicate some of the lines of research which industrial archaeologists are already beginning to explore.

In the first place, it is possible to make inferences from the physical evidence of industrial sites, buildings, refuse tips and so on, about the nature and extent of enterprises, about the date at which they operated, and about their success or failure. Examples could be found in all parts of Britain, but a particularly interesting contemporary case is the fascinating reconstruction of industrial activity which is taking place in the Ironbridge Gorge in Shropshire as a result of sensitive excavation and restoration in this prime industrial conservation area. Blast furnaces, workshops and wharves which had been lost to knowledge have now been firmly restored and the historical record is being amended to take account of them.

Secondly, inferences can be drawn from factory and workshop structures about the arrangement of processes conducted in them, and about the number of workers and their conditions of work. An admirable survey of this type has been conducted by Mrs. Joan Day on the Bristol Brass Industry (see *Bristol Brass—The History of the Industry*, David & Charles, 1973). This industry flourished in the eighteenth and nineteenth centuries in the valley of the Bristol Avon, but faded out in the twentieth century to expire completely in 1927. Mrs. Day has made ample use of the surviving physical evidence of brass works, annealing towers, and associated mills and houses, to create a comprehensive picture of how, where, and when, the industry prospered.

Third, machines and other artefacts provide evidence from which inferences can be drawn about the requirements and provision of power for industrial processes, about the details of the processes themselves, and about the sort of accommodation which would be required to house them. Material of this kind is beginning to emerge from some of the assiduous research of industrial archaeologists on wind-mills, water power, and the stationary steam engine. In one particularly important area of industrial history, the Cromford district on the Derbyshire Derwent, examinations of this sort are making possible a much more precise and impressive account of the complex of Arkwright cotton spinning mills than any that had been previously available from standard text books and documentary evidence alone.

Fourth, inferences from transport relics can illuminate understanding of the nature and importance of particular forms of traffic, and of changes in the direction and volume of such traffic. Research on early artificial waterways, on tramways in mining districts, and on some of our large ports, are examples of this sort of work. In the case of the port of Bristol, a close examination of the physical evidence has proved to be valuable in establishing the successive stages in the development of the harbour facilities. The same sort of exercise could be done for London, Liverpool, Glasgow and Hull, as well as a host of lesser ports.

Fifthly, inferences can be drawn from domestic buildings and the social provision of churches, public houses, bath houses and so on, about the density of population, the conditions of life, and the general standard of living of an industrial community. This can be achieved both in planned community developments such as Robert Owen's New Lanark in Scotland or in the complete town built by Sir Titus Salt around his alpaca mills at Saltaire near Bradford, and in those vast areas of working-class housing in our great industrial cities which have no distinction other than the probable imminence of their own destruction. In all such instances there is physical evidence to be examined and assessed, and an opportunity to establish useful historical information.

These are only some of the more obvious ways in which it is already proving possible to use the physical evidence of past phases of industrialisation to interpret economic, social and technological changes and thus to give a new dimension to our historical understanding of the role of technology in our society. There is tremendous scope here for serious students of Industrial Archaeology to make contributions of value to other disciplines and, in the process, to define its own disciplinary base and thus move towards the sort of academic status already established for cognate studies such as Historical Geography and Urban History. The Association for Industrial Archaeology has been formed to promote, amongst other things, the academic standing of the subject, and its members in all parts of the country will be glad to advise and support activities of this sort. The officers may be contacted through the Ironbridge Gorge Museum Trust, at Church Hill, Ironbridge, Telford, Salop, or through the Centre for the Study of the History of Technology at the University of Bath.

FARMER MAN

ROBERT EMMETT

Workers' Educational Association (Western District) and Tutor/Organizer Swindon and North Wiltshire

The use of very, very simple machines is not unique in man; birds are known to drop stones onto eggs to get at the inside; chimpanzees use branches to retrieve objects which are out of reach. Only mankind has gone beyond this primary tentative stage, and evolved increasingly complex machinery to cope with his increasingly complex way of life.

Unlike animals, man is not evolutionarily adapted to any particular way of life. Though he is a predator, he has not the teeth and claws of the tiger. Though a builder of tunnels and dams, he has not the excavating tools of a beaver. What he has got is a unique brain and adaptable hands, which he has extended by machines to adapt to any particular situation. Now he is totally dependent on them in a way inconceivable to animals.

Man's method of tilling the soil reveals this in the shapes and patterns of fields around us.

Five stages can be clearly seen:

1. *Hand Tilling* by hoe and spade.
2. *Prehistoric Ploughing* with a wooden or antler plough.
3. *Medieval Ploughing* with draught oxen and later horse plough.
4. *Industrial Revolution* by steam plough traction.
5. *Modern agricultural engineering*—the mobile factory.

1. *Hand tilling* resulted in very irregular fields, seen abroad as 'bush fallowing' or 'Chitimene' in Zambia—the burning of brush, then the use of ashes to fertilize the soil with the help of very primitive tools. The tribe then moved on when the soil was exhausted. Some authorities attribute the Sahara Desert to bush fallowing. These effects are not very easily seen in Britain.

2. *Prehistoric ploughing*, using the prehistoric antler or wood plough is so primitive that it has left very little trace on the ground. Fyfield Down is perhaps the best example. Because of the lightness of the plough, it could only be used on very light, thin soils, high on the chalk where traces are found today. As the farmer used better, richer soil in the valleys, he left

these historic relics behind. Man to this stage lived in small hunting communities. Arable agriculture was part of his hunting and gathering existence and he lived in family groups and small tribes. He was not yet a true farmer.

3. *Medieval ploughing* began with the Anglo-Saxons and Danes who brought with them oxen-drawn heavy iron ploughs which enabled man to move the heavy but richer clays and loams. The need of such a large team (eight or ten oxen) aided the setting up of a communal way of life, centred on the village, an example of the co-operative in many senses. The village became a self-contained unit and has left us many signs we can see today, the existing village pattern, ridge and furrow and the open field system. The eight oxen ploughs were very difficult to turn but ploughed well. The result was long strips of ownership, 220 by 22 yards. These strips still exist though they are only worked as museum pieces in Portland, Laxton and Branston but they can still be seen throughout the country.

4. *The Industrial Revolution* brought steel ploughs, steam traction engines, and capitalism. Richer people enclosed their lands. Steel plough shares and the improved technology of the horse harness and collar greatly increased the efficiency of the horse as a source of power. Horse ploughs and the traction engine logically led to square or rectangular fields and the Parliamentary Enclosure Acts. The grid of this scientific revolution can be seen today. It completely swept away the open common field and brought capitalism, the doctrine of a man's getting on or not by his own efforts without recourse to his peers. This resulted in discrete parcels of land or fields, with the farmer living away from the village on his farm. In consequence, there was much farm building 200 years ago and with it came the decline of the village which was only rescued by 'commuters', first through the train and then the motor car.

5. *The modern tractor* needs no rest. When it is turning it is wasting ploughing time. The result is large, rectangular fields. Trees and hedges give place to the prairie as can be seen in Wiltshire and elsewhere.

Can it be a coincidence that, as these technological developments are applied, farm management methods become increasingly industrial, that contract labour and remote boardrooms replace the farmer who lived and worked on his own land, that the short-term productivity of continuous cereal farming replaces the less profitable but possibly greater long-term benefits of crop rotation and mixed farming?

There is much concern that these new methods will ruin the land, and consequently the farmer. Ultimately our civilization cannot survive without cheap food. Probably much of the concern is ill-founded, the ideas being new and therefore suspect. All major changes have beneficial and side-effects, and whilst they are actually taking place, it is difficult to sort out the long-term effects. 'Life can only be understood backwards, but must be lived forwards', sums up this situation well. What the effects will be we do not know. What we can be sure of is that these modern technological changes in farming will be mirrored in economic and social repercussions as they were in the past. The trends are there now but we cannot easily see their effects.

THE TECHNOLOGY OF GAMES

BERTIE DAISH

Formerly Lecturer, Mathematics Department,
Royal Military College of Science

Most people are interested in games. An explanation of many of the effects which are commonly observed in ball games involves the application of a mixed bag of scientific principles, and the field therefore provides a great opportunity for a teacher to present and to illumine his subject in a way which will arouse and maintain the interest of pupils some of whom might well be unattracted or even bored by a more orthodox approach.

In particular, of course, the principles which are taught in a school mechanics course are nearly all capable of illustration and explanation in terms of situations which occur on the playing field. Most ball games involve impacts between the ball and an implement of some kind: a bat, a racket or a club, or the foot or hand of the player. Such impacts provide interesting examples of Newton's laws, the conservation of momentum and of angular momentum, energy conversions and so on.

Most of the impacts involved are of short duration—measurements of impact times can be made using electronic timers which are now in reasonably general supply—and the application of basic principles often leads to conclusions which are not anticipated by the player. In a golf drive, for example, the ball and club are in contact for only about half a millisecond, in which time the head of the club travels little more than a centimetre at the bottom of its swing. After this the ball is flying on its way and there is nothing the player can do to affect its subsequent fate. The short impact time implies a very rapid acceleration of the ball—over 10,000 g in this case—and hence very large forces indeed. Much the same applies in a cricket shot or a football kick. The follow-through cannot, in itself, affect the shot in any way. Then, is the games coach right to insist on the necessity for a full follow-through? He may well be.

The biologist will encounter interesting problems relating to the power which can be derived from human muscles and how the power available in the various parts of the body can best be utilised by the player to achieve the desired results.

In many cases after impact the ball bounces on or rolls along a surface. Its behaviour, especially if it is spinning, can be quite complicated. Effects of this kind have long been dear to the hearts of applied mathematicians, from Coriolis onwards.

Or it may fly though the air and display effects which need, for their elucidation, some fairly simple aerodynamics. Papers by both Rayleigh and J. J. Thomson on this aspect of the subject appear in the scientific literature.

But perhaps the most useful aspect of this field to the imaginative teacher lies in the opportunities it presents to devise group project studies for his students to tackle. Scientific investigation of ball games is scanty, yet every game has associated with it a host of 'theories' which have been built up over the years. These have usually originated from eminent players who have sought to analyse the impressions they experience when executing their expertise and to extrapolate these feelings into 'theories' intended to help lesser players to improve their game. Some of these ideas are correct in that they are confirmed by the results of independent objective observations. Most contain some elements of truth. Some are just plain nonsense. There are so many ideas in the field of games which cry out for a proper assessment, based on measurement rather than on prejudice, and many of these could be examined by a team of pupils with an interest in the game involved.

'A cricket ball swings more in a heavy, humid atmosphere.' This is a statement often made by quite experienced players and commentators. Yet it cannot be correct as it stands because, under the same conditions of temperature and pressure, water vapour is appreciably less dense than air, so that a mixture of the two to form a humid atmosphere must be 'light' rather than 'heavy'. So does the ball in fact swing more in a humid atmosphere? Most cricketers are convinced that it does. If so, why? Nobody really knows.

Does a table-tennis bat covered with sponge rubber impart significantly more spin to the ball than one with the old-fashioned sandpaper or thin rubber surface? At first sight this might appear self-evident, but a consideration of the effects of friction in imparting spin makes it less likely. So does it really happen? If so, why?

It has been suggested that a table-tennis table with a three-ply top gives a higher bounce than one with a solid top and even that a table on a wooden floor gives a higher bounce than if the floor is covered with lino. Is there anything in this? If so, where does the explanation lie?

How does the weight of a cricket bat affect the energy the wielder can impart to the ball? Or the weight of a hockey stick, or a tennis racket? Or the tension in the strings of the racket?

What advantage does a big, beefy player possess over a weaker opponent in imparting power into his shots? I have a sneaking suspicion that, in popular games, this advantage is not as great as one might imagine—mismatched impedances, perhaps. But am I right?

One could go on posing such questions for a very long time. All of them seem to be interesting to me. Any group tackling one of them would learn a lot about scientific method. They might well be able to help their fellows to derive more enjoyment from their leisure-time activities and their findings could well interest the manufacturer of sports gear or the administrators of the game. Many of these gentlemen have still a lot to learn.

A gentle word of warning. Most problems in this field involve a player making a shot or performing some kindred action. Now one of the fascinations of a good game is that not all such efforts are equally successful and few are perfect. We all display what the artillery gunner refers to as large 'round-to-round variations'. So a large number of similar observations are necessary if any real facts are to emerge. How many in a particular case is an interesting statistical calculation. Similarly the results must be subjected to a statistical appraisal to sift any real truth from random fluctuations. Much prayer and fasting may be called for. But this is true in so many scientific investigations and, with a group of enthusiastic workers, large numbers of observations can be made reasonably quickly, often with the aid of quite simple apparatus. And the statistics involved is fairly simple.

Where to begin? Somewhat embarrassingly I would recommend anyone interested to get hold of my book—*The Physics of Ball Games*, published by English Universities Press—which touches on some of the problems and gives a bibliography. If you think this is blatant advertising—and I am conscious that it reads a bit like that—do not encourage me in it. Borrow a copy from a library.

Purpose

THE CHRISTIAN VIEWPOINT

THE RT. REV. JOHN S. HABGOOD

Bishop of Durham

> 'Would you tell me please,' asked Alice in Wonderland, 'which way I ought to go from here?'
> 'That depends a good deal on where you want to get to,' said the Cheshire cat.
> 'I don't much care where—' said Alice.
> 'Then it doesn't matter which way you go,' said the cat.

The point is so obvious that it scarcely seems worth making. A society which does not care about ends and purposes has no means of knowing whether so-called technological advance is good, bad or indifferent; or indeed whether it is proper to call it 'advance' at all.

Yet things are not quite so simple as the Cheshire cat supposed. Choices about the way to go may be greatly influenced by the direction in which we are actually travelling. Technology has a flywheel effect. There is a momentum in it, as there is in science, pushing society towards goals which very few people may want, but which are felt to be somehow inevitable. Thus because it was technically possible to develop supersonic passenger aircraft, the odds were stacked in favour of building them, despite the grave doubts about whether they were really needed.

The flywheel effect can only be resisted by those whose convictions are strong enough to enable them consciously to choose their goals and stick to them, instead of having them forced upon them. This is one reason why a positive religious faith can be important. In our Western society Christianity provides a point of reference outside the technological process—a source of awkward questions.

Can we, for instance, separate technology from social justice? In a world where the greatest needs of the vast majority are for the basic necessities of life, how far is it right to pursue a policy of unlimited growth and development? In a world where resources are limited, ought we to be searching for the maximum sustainable quality of life in which all can share, instead of developing technologies which presuppose a very sophisticated, and possibly short-lived, style of life for a minority?

What are genuinely human ends? and are real human needs satisfied through greater and greater consumption? Or is there a point at which freedom from drudgery becomes enslavement to machinery? High-cost industrial machinery, for instance, can create enormous and sometimes damaging social repercussions. The large capital outlay has to be paid for by having machines in continuous operation, which in turn involves shift work.

How does one distinguish between problems which are worth solving, and problems which have simply been generated by the desire for novelty and the need to stimulate demand? Ethical questions are now being asked about the use of animals in testing new cosmetics, most of which achieve little more than those they replace, though hundreds of animals may have to be sacrificed in satisfying toxicity requirements.

Questions about the proper use of human skills have recently been posed in a very practical way by a Shop Stewards Committee in Lucas Aerospace. They have argued that the skills developed in the aircraft industry could be diverted to humble and useful ends, such as the provision of artificial limbs; and they have a simple device, designed for children with spina bifida, to prove their point. It is true that one of the motives in this instance is fear of redundancy, but their general claim is a valid one. 'We have allowed our regard for human talents to be bludgeoned into silence by the mystique of advanced equipment and technique, and so forget that our most precious asset is the creative and productive power of our people.'

This reference to human creativity can serve to introduce the positive side of what a Christian would want to say about the purpose of technology. When all the awkward questions have been asked, there remains an inherent value and fascination in inventing things. It is part of man's task, as understood in theological terms, to master the natural world and use it for the glory of God. Man is made in the image of God the

Creator. Technology is thus to be accepted gratefully as one of the ways in which human beings find fulfilment, and human life is enriched. Nevertheless because man is man, and not God, he needs to be aware of his limitations and ignorance, as well as his powers.

Technological blunders occur, as in any other sphere of life, and the cure for them may be greater watchfulness. Failures caused by not paying sufficient attention to the complexity of the natural or human environment call for a rather different attitude of mind, something nearer humility. The nutrition experts who undertook an expensive research and development programme to turn anchovies into an odourless, tasteless fish flour, omitted to discover from the protein-deficient people for whom it was intended whether they preferred anchovies—and found, too late, that they did. This was a failure to recognize that they were dealing with real human beings, rather than statistics of nutritional deficiency.

The unexpressed assumption that human problems ought to have technological solutions is another example of a failure of humility. It is very tempting for those who have learnt to use powerful problem-solving techniques, to assume that they can know enough about the complexities of human life to apply these techniques without substantial risk of error. But technological solutions to the problem of war, or even of riot control, for instance, have not been a conspicuous success. And it is a sad thought that the crudest example of looking for a technological fix to meet a human need is to be found in the drug culture.

A Christian approach to purpose in the technological society must endeavour somehow to do justice to the ambiguity of technology. It is a tool in the hand of man, which can be used for good or ill. There is no absolute value in technological advance, and this is why questions have to be asked, which raise much wider social, political and religious issues. Such questions are especially urgent when the tool seems to be dictating to the man, rather than vice versa.

Yet on the other hand, there are no good grounds within the Christian faith for despising or failing to appreciate the enormous achievements of technology. A sacramental view of life should lead to a reverence for material things which, like the sacraments themselves, can become a means to spiritual ends. The need is to find ways of using our skills and our creations for purposes which enhance the lives of all. And this entails both humility and hopefulness.

Humility is necessary, because the Christian sees himself as holding the world in trust from God. It is not his to exploit unmercifully, but to use with respect. Hopefulness is possible, and can motivate human effort, because in Christian belief the universe is not ultimately hostile or meaningless. Human achievement can have an abiding significance.

Further Reading

Some of the most interesting and comprehensive material on this theme is produced by the World Council of Churches. An occasional paper, called *Anticipation*, provides information about a continuing ecumenical enquiry on 'The Future of Man and Society in a World of Science-based Technology'.

Other papers on the same theme are published in *Study Encounter* (e.g. SE/50 on the problems of economic growth). Obtainable from Publications Office, W.C.C., 150 route de Ferney, CH-1211, Geneva 20.

THE HUMANIST CASE

HARRY STOPES-ROE

Senior Lecturer Science Studies, Department of Extramural Studies, University of Birmingham. Chairman, British Humanist Association

Though technology (rightly used) and Humanism are twins, the abuse of technology, which now troubles so many thinking people, is the result of society's past and present failure to come to terms with the reality of fact and value. This the Humanist sees most clearly: the claim, of course, is much oversimplified, but Humanists see a core of truth in it. How can we manage man's creative power, unless our hearts and minds both turn first to our fellow men, and unless we acknowledge that man alone is responsible for the power he creates, and the guidance of it?

It is sometimes said that Humanism has no 'absolutes', but this is a misleading oversimplification. One Humanist 'absolute' is clear and simple in intent: human satisfaction. Such satisfaction, fulfilment, happiness, or call it what you will, is of supreme value and importance. The word 'happiness' is often debased, but I will use it. The Humanist conception of 'human happiness' is by no means the same as, for example, the Christian—but it is the base of Humanist purpose in technological society. But happiness must not be limited in its extent, either over individual people or over time, and this leads us to another Humanist 'absolute': reason, working with sympathy and insight, on evidence. How else can we extend the range of happiness? This question is a serious one, and the Humanist answer to it is: reason, sympathy, insight, (natural) evidence are *all* we have. If they are not enough, nothing better can be found. Just as we cannot justify our actions by appealing to happiness in an afterlife, so we cannot find guidance or assistance for our actions in a transcendent. There is no past revelation, present guidance or future support from a transcendent. If we want to make things, or improve our circumstances, we must do it ourselves. *Humanist* Technology is ordained explicitly to the free happiness of men (without reference to any non-natural state); and it is created by reason and sympathy, using all the evidence relating to human and material conditions. Finally, it is Humanist because it is directed by men—without reference to any Transcendent, either for guidance or for assistance. *All* problems are laid at *our* door.

This might, at first glance, look rather like the technology we experience in Britain all around us now. Certainly, it is a fact that technology has become detached from all transcendent reference, apart from such minority practices as the blessing of motor cycles and the healings at Lourdes (medicine is a technology). But 'western' technology and the technology conceived above as Humanist, differ in their purposes, and hence in their reality.

A crude feature of modern 'technological' society which is condemned by Humanist and Christian alike is its exploitative and manipulative quality. Another criticism which, I think, expresses a similar point is that technology is treated as an end in itself. Too many people have no scruples when they act upon people to secure their own advantage; and this is often mediated through some 'technological' device or method. The device or method is advertised as if it had a value in itself, or value just because it is a technical advance. But the technological gimmick is merely the front for the advancement of the sponsor, not the end in itself. Reference to such advertising is a fruitful way of entering the discussion of what is wrong with technology as now practised. One must notice, however, that there is a slightly subtle point here: there is a human satisfaction in owning something that is 'better' in some sense.

Even where technology is applied with good will, the result all too often is not human happiness, because the analysis is inadequate. We may be tempted to feel intolerant of such failure, for it leads not only to an immediate bad outcome, but also to discredit for the technological ideal. But powerful technology is so new, and the achievements are so exciting, that it is not surprising that practitioners fail to appreciate the range and subtlety of the human factors that mutely demand attention—or are vociferous in inappropriate ways! In particular, so many factors are invisible, and so many significant outcomes are unpredictable, and so often long-term and short-term conflict. Planning is a fruitful source of examples here, and transport. Gas and oil illustrate the long/short-term conflict. Pesticides illustrate the unexpected outcome— immune varieties, ecological interactions, etc. All these examples require care in use, for though sometimes vigorous condemnation is required (the promotion of infant foods so that they replace breast feeding in underdeveloped countries, for example), more usually the technologists are sincerely doing their best. They deserve sympathetic understanding. And without that, we will not understand the *difficulties*, and then no advancement can be expected.

The assessment of technological means for human ends requires one to assess situations that have a complexity far beyond ones techniques. This is a typical 'engineering' situation, but it is engineering of a new kind, for the assessments are applied to human qualities and feelings, and hence require human sympathy.

The most serious present lack is concern for individual happiness. The old transcendentalist foundation for values has crumbled, along with the superstructure of belief in God. Such residue of the protestant ethic as survives is a respect for hard work, and for exploitation. Our present climate is directly inimical to the true purpose of technology—happiness —and also it opens the door to the manipulation of people's desires: for many people now 'Happiness is an AVENGER car'—and the vicious circle takes a tighter twist. 'Consumers' can be manipulated so that they accept unrealistic satisfaction—a 'better' car offers, not safety, but a power of self-assertion.

Present technological society does not succeed. We have three courses to choose from. We can seek the guidance and power of God. We can opt out of the task, saying that modern technology gives such power to man that he cannot cope. We can set to work on the technologists and on those who control them, and try to develop two sensitivities: a sense of their total responsibility, and a vital sense of the presence of a vast multiplicity of individual human beings. These are the principal sensitivities, but there is a third also: to our humanity as a part of nature, guided only by natural evidence. We are finite, and so must act cautiously. It is of the essence of scientific understanding that it is partial and approximate.

How can we afford to despair, if we can do something, and if this world is all we have, and we alone are responsible for it; and if men and women press around each of us in their sufferings and joys? Can we suppose that we would do better with obscurity and false hopes?

THE EXPERTNESS OF EXPERTS

JEROME RAVETZ

Reader in the History and Philosophy of Science, University of Leeds and Executive Secretary, Council for Science and Society

The Problem

Where have all the experts gone? Those distinguished men, with such authoritative voices, who assured us that computers would soon do all our drudgery; that Concorde would fly us to New York for lunch and back for tea; that the Green Revolution would feed the world's hungry people; that urban motorways are the essential civic monument of today; and that nuclear power stations have such cheap fuel that they'll give us energy in abundance.

Yes, they're gone, in a sea of worries and confusions. Instead of their smooth-shining optimism, we have new crises battering us from every quarter. Yet another sort of insidious poisoning on the street or in the factory, that *could* have been detected if anyone really cared; yet another energy worry, as nuclear power stations and North Sea rigs alike run into construction difficulties. In the mounting tale of planning muddle and technical incompetence, we discover that no one is in control, and probably never has been.

So, the future, or our vision of it is in question as never before. We can start to ask seriously *why* we want this or that, and not be fobbed off by magic letters like 'GNP'. It is even possible for someone to write a book on economics 'as if people mattered'. No longer can all the questions of the quality of life be reduced to quantities in guesstimates carried out to three significant digits.

One of the greatest technological experts of them all has invented the word that is the passport to liberty in technology. The word is 'trans-science'; its author is Alvin Weinberg, until recently the head of the American Atomic Energy Commission's Oak Ridge Laboratories. He realised that some problems of safety in nuclear engineering cannot be decided scientifically, even though their statement is quite scientific. His example was the radiation damage from nuclear pollution releasing radioactivity only as intense as natural background radiation. This *could* be tested experimentally, but it would need about a thousand million mice! Then other long-term pollution problems, as the release of nitrogen oxides into the stratosphere by a fleet of Concordes, depend on simplified models with a very thin data-base; arguing either for *or* against a hazard becomes speculative.

In situations of 'trans-science', the public and the policy-makers cannot turn the matter over to the experts. The experts can advise, but their views will always be tentative and controversial. So the final decision on a 'trans-science' policy question must be political, which involves the art of weighing imponderables and arguing by intuition.

Some will say that 'everyone knows' it has been like that all along. It may well have been so, but it has been an extremely well-kept secret until recently. The discovery of trans-science in *one* place will soon be followed by its discovery everywhere. If you ask for the justification of *any* formula used to define a 'safe' level, or an 'acceptable' risk, you will see how little the scientific part of the argument determines the answer. To take a simple case, why is the urban speed limit 30 rather than 25 or 20, as elsewhere? Are no children killed by the faster cars, or are they negligibly few? From enquiries such as these might come truly educational class projects in technology.

How We Got Here

Ours is a civilisation of inventors. Unstable, aggressive, intolerant, Europe has always been thrusting forwards and outwards; this goes back to medieval times. This is our fate, unlike the great Eastern civilisations that value social stability or contemplation over the fruits of conquest of men or Nature.

In Europe the inventor, the engineer has been an important figure; the successful ones could be patronised by the mighty, and rise far above their station. The streams of inventions and their ruthless exploitation first brought Europe abreast of the more advanced civilisations of its Islamic neighbours, then took it out to conquest and genocide in the New World, and ultimately created the manufacturing system from which unimagined wealth and power have flowed. The Victorian cities, rather like Stalin's Russia, were heaven or hell, depending on your point of view. To the wealthy they were fountains of commodities benefiting all mankind while enriching the masters. To the operatives, driven from rural starvation into wage slavery, the benefits were slow in appearing.

With the maturing of the industrial system, children being protected from destructive exploitation so that they could at least survive to breed, and soap and sewers making the cities fit for animals to live in, came an awareness of more refined social goals than simple 'muck and brass'. Our century has seen the rise of social welfare as an important criterion of technological progress. Citizens should be well fed, clean, literate and upwardly mobile.

Welfare, of any sort, involves an interference with the effects of the cupidity that makes men work; and this is done through a bureaucracy. The modern State and Corporation, whatever the property system or political structure, is inevitably bureaucratic. Things are then run by committees of faceless men; more smoothly or humanely (usually) but with less excitement, and perhaps less humanity. Also, whereas previous possessors of power argued for themselves on the basis of inheritance or wealth, the bureaucrats claim to benevolence and expertise.

Without public faith in the expertness of our experts, modern society could not be managed. From the setting of the Budget and the Bank Rate, down to school syllabuses and exam questions, decisions made by bureaucrats provide the constraints and channels for our activities. If people lose belief in their competence and legitimacy, they are lost—there are never enough policemen to enforce a really unpopular decree. And among the most destabilising elements of our present very unstable situation is the discrediting of the technobureaucrats, those who claim to engineer nature and society on our behalf.

What Next?

In some ways this new demythologising can be a dangerous thing. There is a guaranteed minimum of competence and probity in work, when it is done by someone with a certificate; if left to the amateur or enthusiast, the performance can be quite astonishingly bad. The defenders of the *status quo* always put the rhetorical question, would you not have surgeons trained and licensed?

Yet the game to be played by technology, techniques and expertise in our society, is changing radically in spite of all defences. The question is whether the schools, themselves, conservative institutions in some very important respects, can contribute to making this change a healthy one. I shall resist the temptation to sketch out grand schemes and syllabuses, since that is the same oldfashioned heavy engineering approach that has become so destructive, in the schools or elsewhere.

Let us think about projects; and try to design them so as to be less prone to abuse by lazy teachers and apathetic pupils. Perhaps some of the things that people do with their hands could be fun, and educational.

Much of the 'alternative technology' now being developed is small scale, inexpensive and hand-made. A windmill project, generating electricity or pumping water, could be an exercise in 'junk technology', and in energy economics. A methane digester could be a useful case study of how things can *nearly* work, but don't. Pollution studies, associated with local industrial establishments, might lead to political delicacies, but could be all the more instructive thereby. An urban garden, with plants more or less resistant to acid, or with lead-levels checked on grown specimens, could make food problems more real.

Of course, there are discussion topics too, about Energy Policy, Radioactive Wastes, Food Imports and many more. Proper discussion of these could be a stimulating experience for science teachers, for now these are (as I have indicated) as inconclusive as the supposedly 'soft' social-studies topics of The Family, Social Class, Equality and so on.

We need not take all this too seriously (and it is part of the new style of doing things, never to take ourselves too seriously), since 'the media' are covering these topics, and forming young people's attitudes and expectations quite effectively already. It would be enough if schools concentrate on what is uniquely theirs; the imparting of useful skills, and assisting towards intellectual discipline.

Out of all these social and educational changes will come something different—better in some ways and worse in others. But if we succeed in developing a healthy critical attitude among young people towards those professing a monopoly of expertise to do good on their behalf, the world will be a bit safer, and more interesting as well.

THE ECOLOGY OF DESIGN

KEN BAYNES

Research Fellow, Department of Design Research, Royal College of Art

Why design-education? Is it to revitalise a stagnant economy; perhaps to inform 'public opinion' so that engineers, architects and designers can work more effectively? Or, is it to give them the knowledge and skills to control and direct the experts to create a world which is more responsive to the needs and wishes of ordinary people?

It is to the economist Adam Smith, who died in 1790, that we have to look for one of the first convincing codifications of the designer/technologist's role in industrial society. Smith summed up the significance of the economic and technological forces which had been gathering strength since medieval times. In his treatment of mass production and the division of labour he foreshadowed with complete accuracy the emergence of the designer as a part of the modern world of highly educated specialists. Here, in a single quote from a draft of his famous book *The Wealth of Nations*, is the key:

> In opulent and commercial societies ... to think or to reason comes to be, like every other employment, a particular business, which is carried on by a very few people, who furnish the public with all the thought and reason possessed by the vast multitudes that labour.

Such a high level of generalisation was founded on accurate knowledge of efficient industrial practice. The tendency to specialize which he identified was dramatically effective and destroyed once and for all the old work and craft relations in society. The salient point for educationists is to recognise how rigid and impervious to equality and liberalism is any such vertically organised specialisation. The division of labour does not imply the division of responsibility and power: quite the reverse.

There is, however, the other, triumphant side to the division of labour. The *Wealth of Nations* emphasised what it meant to have hopes of a growth in human prosperity, of a spreading of the enlightenment and a consequent extension of civilisation. Smith predicted with satisfaction that the division of labour was a radical element in society. He saw that a vast increase in productivity, based on technology and modern methods of production, must inevitably change the power relations in society. Smith looked to a time when even the labourer in a civilised community would be better accommodated than 'many an African King, the absolute master of the lives and liberties of ten thousand naked savages'!

Smith argued that when shortage disappeared it would be open to men to extend their other capacities in exactly the same way as the division of labour was already extending their productive capacity.

Any hungry man would agree with Smith about the deforming and destructive aspects of shortage. What is less easy to evaluate is how far the structural inflexibility of industrialisation has been such as to negate the advances that might otherwise be expected from the spread of wealth.

Smith had little to offer on such pressing contemporary problems as pollution and resources, but he did encompass the precise dimensions of the problem of political power facing industrial man. Already, in 1790, he recognised the trade off that the consumer and citizen still make today between plenty and involvement, and he identified the interplay between 'bad' work and extended leisure.

Smith spelt out the emergence of a social contract based on the existence of boring work and the compensations of vastly extended consumption and entertainment. He penetrated to the core of the modern paradox. The division of labour brought a dramatic increase in economic power for the mass of the people. In this respect it gave more power to the community than before. But with it went the emergence of bureaucratic specialists, many of them designers of one kind or another, who made decisions on behalf of society and, therefore, controlled its development. In this respect, the community had, and still has, extraordinarily little power to affect very large and important areas of decision-making.

The question for those engaged in design and technological education is: should they deliberately set out to redress this balance? And, what would be the likely effects of such an activity?

Perhaps, what we need is something more like an 'ecology' of designing. The use of the concept of a 'solution' implies a moment of stasis when strains are resolved and the new way is set. This is not a moment which exists in any ecological situation and, in design, it is really only a convenient but confusing fiction. In ecological models we have continuing processes and interactions which never cease acting on each other and which never stop. The same is true of those systems of which design is a part.

It will help to take an example.

It is frequently said that planners fail because 'they do not know what people want'. They do not solve people's problems. But most 'bad' designs do not arise because designers are ignorant of what people 'like' or 'need'. They arise because, on a completely different level, one group of people is in conflict with another. What may express itself as a conflict between planners and users is in reality nearly always a conflict between third parties in which *any* solution is to the gain of some and the loss of others. Consider the vaying degrees of power wielded by the parties to a redevelopment, and also their varying degrees of access to the designer whose job it is to find the means to ends that are inevitably incompatible.

In any such example it is significant that it will be those groups who *already have power*, who are also the ones with access to the designer. It is they who will pay him and give him his brief. It is also significant that, in direct human terms, it is not they who will be the people affected most intimately. It is not they who will eventually have to move after having had the experience of living for a large slice of their lives in a steadily deteriorating slum fatally affected by 'planning blight'. What we are looking at here is not in any sense describable as a 'problem' which the designer could 'solve' by his design. The designer, in his design, will inevitably be throwing the weight of his skill onto one side or other of a group of conflicting interests. Where this kind of intractable social conflict involves design decisions, it is interesting to recognise how far it differs from the situation which we expect to find either in science or in the majority of narrowly defined technological designs.

The differences between them have been well described by Horst Rittel and Melvin Webber* who have given these 'problems of appreciation' a splendidly apt name: they call them 'wicked problems'. Here is an extract from what they have to say about them:

> The kinds of problems that planners deal with—societal (*sic*) problems—are inherently different from the problems that scientists and perhaps some classes of engineers deal with. Planning problems are inherently wicked.
>
> As distinguished from problems in the natural sciences, which are definable and separable and may have solutions that are findable, the problems of governmental planning—and especially those of social or policy planning—are ill-defined: and they rely on elusive political judgement for resolution. (Not 'solution'. Social problems are never solved. At best they are only re-solved—over and over again)....
>
> On the other hand, the problems that scientists and engineers have usually focused upon are mostly 'tame' or 'benign' ones. As an example, consider a problem of mathematics, such as solving an equation; or the task of an organic chemist in analysing the structure of some unknown compound; or that of the chessplayer attempting to accomplish checkmate in five moves. For each the mission is clear. It is clear, in turn, whether or not the problems have been solved.

They go on to say, 'The lay customers are complaining because planners and professionals have not succeeded in solving the problems they claimed they could solve. We shall want to suggest that the social professions were misled somewhere along the line into assuming they could be applied scientists—that they could solve problems in the ways scientists can solve their sorts of problems. The error has been a serious one.' Here is an insight for teachers of design and technology. The most important design problems are 'wicked' problems of 'appreciation'. They have no clear-cut beginning and no clear-cut end: they involve morals and politics as much as technology. And they are too important to be left to professionals like designers: they are about issues which the community *as a whole* must think out for itself.

*Horst Rittel and Melvin Webber, 'Dilemmas in a general theory of planning', *Policy Sciences*, **4** (1973).

The International Dimension

EUROPEAN SCIENCE AND TECHNOLOGY

ARTHUR HOWCROFT

Department of European Studies and Modern Languages, UMIST

Science and technology as they are now practised are both worldwide and international in character and scope. For scientists the universality of scientific theories, principles and discoveries is so evidently true and sacrosanct that any attempts by nations or organisations to prevent the dissemination of such truths is met with loud cries of protest and dissent. Technologists, more often placed under practical or economic restraints and sometimes with an understandable pride and belief in the quality of their product, may appear to be more nationalistic and less internationally minded. Even so, technologists use by and large the same methods of manufacture wherever they happen to work, be it in Brest-litovsk or Birmingham, Japan or Jakarta. Moreover, international projects carried out by technologists from many different nations are now commonplace. Technologies then can also claim, apart from differences in detail brought about by local social and physical conditions, a kind of universality in method and end product that is not restricted to a particular area.

How in this context can one talk about *European* science and technology? Except in the comparatively restricted sense of 'European' projects or in the sense of studying what a particular European nation is doing the idea seems to have little meaning or significance. But as soon as one considers the idea in the sense of the origins, development and diffusion of modern science and technology then it immediately takes on a far greater significance and scope. For there can be little doubt that modern science and technology originated and developed in Europe, although of course much that was significant and important—the compass and printing are obvious examples—came from elsewhere. It was in Europe that scientific method as we now understand it originated and in Europe that the development and application of inventions and discoveries, whatever their origin, took place and became important. Indeed it can be argued that, together with Christianity and democracy, science and technology have been the most powerful forces at work in shaping not only Europe itself and giving it its present life-style and generally high standard of living but also in spreading this life-style and standard of living or aspirations towards them throughout the rest of the world. A theme as large as this, offering so many possibilities, cries out for a central place in European or Integrated Studies syllabuses yet it rarely if ever appears.

The reasons for this are not far to seek. Most teachers involved in the teaching of European or Integrated Studies have received little or no training in any science or technology. Half-remembered 'O' level Physics or Chemistry only serves to reinforce the general sense of inadequacy to cope that ensues even if one admits the importance of the theme and would like to use it. In any case, science and technology appear to be divided and subdivided into so many complex specialisations that even the science teacher trained in one branch is often reluctant to attempt to teach another. The best—perhaps the only—means of overcoming these difficulties is by team-teaching. In this one would almost be teaching by example since no scientist or technologist would embark on such a project alone and would accept as normal the co-operation and assistance of others.

Given that such co-operation and assistance exist, how could a course in European Science and Technology be developed? Two major contexts suggest themselves: space and time. If it is accepted that modern science and technology are European in origin then it would surely be worthwhile to examine them first in their place of origin, i.e. Europe. This could be followed by an examination of European science and technology at work outside Europe. From the comparatively simple 'who did or does what where' to highly sophisticated problems concerning the interactions of science and technology and different cultures, this space context already teems with topics, and themes at almost any level. If to these a time context in the shape of some history of science and technology is added, then the course becomes one of truly enormous variety and scope. Teachers of geography, history, science and technology, working together, should have little difficulty in finding a host of topics and themes to develop. The language teacher too may have much to contribute since many major pieces of scientific writing have first appeared in languages other than English, notably French and German, and await exploration. Contrary to the belief of many language teachers, they are not all as difficult as might appear and are often very well written.

The presentation of science and technology in a historical context has many advantages especially for the non-scientist. Many of the ideas and theories which were once startlingly new and too advanced for general acceptance (the heliocentric theory and evolution for example) are now common knowledge and therefore more easily understood. Most major developments in science and technology have occurred in an international context and tracing such developments historically soon reveals how truly European they are.

But for non-specialists, however willing they may be to incorporate these suggestions into their teaching, the problems of dealing with specialised branches of science and technology and the resulting mass of complex information remain a deterrent. In this context, teaching the history of one specialised branch, say physics or chemistry, is not really a solution since time and expertise are all too likely to be in short supply. An overall picture of European science and technology is also very unlikely to emerge. What is required is a more generalised history and fortunately the actual developments that took place in European science and technology can be used as a useful framework here. For when one looks at the overall picture of European scientific and technological developments, what is striking is the way in which at particular periods certain preoccupations and problems seem to dominate. They result in each case in a recasting of some vital aspect of our ideas regarding natural phenomena. This 'break-through period' is usually comparatively short although the period of gestation when the problem is dominant may be long, and it is followed by a fairly long period of consolidation when the implications of new theories, principles and laws are worked out and tested.

The Scientific Revolution of the sixteenth and seventeenth centuries in astronomy and physics is usually taken to be the first such recognisable period. Involving as it did, Copernicus, Brahe, Kepler, Galileo, Descartes and Newton, it set a European pattern that was to be repeated more than once. One or two nations might well be dominant for a time but none could ever claim a monopoly.

In the eighteenth century a new gestation period began, this time for chemistry. The outcome in the late eighteenth and early nineteenth century was a new set of ideas concerning the ways in which substances combine with others. The names of even some of the principal scientists involved (Priestley, Lavoisier, Dalton, Davy, Avogadro, Berzelius, Liebig) indicate yet again how truly European these developments were. This period also sees the rapid 'break-throughs' in technology generally referred to as the Industrial Revolution. In Britain, it is usual to present this as a phenomenon taking place mainly in Great Britain and hardly anywhere else. This is, of course, not true and an examination of what happened in other European countries, especially France, the Low Countries, and Germany opens up a whole host of possible topics concerning the diffusion of science and technology via men, ideas, inventions and industries, both in Europe and elsewhere. Equally numerous themes and topics can be found in the gradual fusion of science and technology which 'accelerates' at this time as it continues to do.

Later in the nineteenth century, work in the life and earth sciences and on the nature of electricity lead to further revolutions in the form of the Theories of Evolution, of the Structure of Matter and of Relativity. Again the stage is held by scientists from many different European countries. In the technologies, the diffusion process via 'second generation', i.e. emigrant Europeans, really takes off and the impact of European science and technology on and its adaptations to other parts of the world from that time to this make for manifold themes and topics in themselves.

This brings us to our own time and present preoccupations with the environmental sciences, nuclear physics, the resources crisis and our current re-evaluation of the kinds of science and technology we want or can afford.

As already suggested at the beginning of this article, these activities and the resulting preoccupations are now worldwide. But it must be emphasised that the implantation or absorption of European science and technology, no matter where it comes from geographically, still tends to 'Europeanize' the host cultures. For future Europeans, therefore, some knowledge and awareness of their science and technology and its role in the world ought to become a vital part of their education, if they are to solve some very large problems they will face. It is unfair and no longer enough to leave the working out of solutions to the scientists and technologists. In any case their specialisms sometimes prevent them from seeing the wood for the trees. So we must all try to explore the forest no matter how difficult the route-finding may prove. A course in European science and technology built on the framework suggested above could be the first nature trail that leads to exciting discoveries, a more confident familiarity with the modern landscape and above all vital knowledge with which to face the future.

Further reading

Bernal, J. D., *Science in History*, 4 vols, Watts, 1969 (Paperback Penguin, 1969).
Butterfield, H., *Origins of Modern Science 1300–1800*, Bell, 1957.
Cardwell, D. S. L. C., *Technology, Science and History*, Heinemann (Paperback), 1972.
Crowther, J. G., *A Short History of Science*, Methuen, 1969.
Daumas, M. (ed), *Histoire de la Science. Encyclopédie de la Pléiade*, vol., Paris, Gallimard, 1958; *Histoire générale des techniques*, 4 vols., P.U.F., 1962.
Derry, T. K. and Williams, Trevor, J., *A Short History of Technology*, Oxford, OUP (Paperback), 1970.
Fortes, R. J. and Dijksterhuis, E. J., *A History of Science and Technology*, Pelican, 1963.
Hall, A. R., *From Galileo to Newton*, Collins, 1963; Penguin paperback *The Scientific Revolution 1500–1800*, Longmans, 1962.
Klemm, F., *Technik: eine Geschichte ihrer Probleme*—Karl Alber Verlag, 1954; Trans. *A History of W. Technology*, Allen & Unwin, 1959.
Kranzberg, M. and Pursell, C. W. (eds.), *Technology in W. Civilization*, 2 vols., OUP 1967.
Mackenzie, A. E. A., *Major Achievements of Science*, 2 vols. Vol. 1 *History*; Vol. 2 *Texts*, C.U.P., 1960.
Mumford, L., *Technics and Civilization*, Routledge, 1934; The Myth of the Machine, Secker & Warburg, 1967.
Taton, R., *General History of the Sciences*, 4 vols. (trans. from French), Thames & Hudson, 1964–6.
Tovlmin, S. and Goodfield, J., *The Architecture of Matter*, 1962; *The Discovery of Time*, 1965; *The Fabric of the Heavens*, 1961; Hutchinson & Penguin paperbacks.
Woodruff, W., *Impact of Western Man*, Papermac, 1966.

THE THIRD WORLD

ARNOLD PACEY

Appropriate Technology Officer, OXFAM

All the countries of Asia, Africa and Latin America, that comprise the 'Third World', have certain problems in common. One of them is the basic question of whether their peoples—500 millions of them—get enough to eat.

Insufficient food (undernourishment) and the wrong kind of food (malnourishment) are probably the most common causes of illness in these countries. Food deficiencies cause many children to die in infancy, and of those who survive, many become blind or suffer permanent brain damage through lack of adequate nutrition. Many people cannot do a full day's work because their food does not provide sufficient energy.

This situation has worsened in the last 30 years, because populations have been increasing rapidly throughout the Third World, but the development of agriculture has been much slower. Many countries which had food surpluses for export in 1940 now suffer from food shortages and can ill-afford the imports necessary to feed their peoples. And some countries have actually lost some of their potential for food production as a result of damage to the environment—destruction of forests and bad farming methods have led to the extension of deserts and to soil erosion.

One purpose, indeed one priority of the technological society, ought to be to ensure that these problems are solved and that the world will be able to feed itself in the years to come. Progress in reducing the amount of hunger in the world is not only necessary in itself, but because of what it can contribute to the solution of so many other problems as well. Better standards of nutrition would lead to better child health and fewer deaths among the very young. With more of their children surviving, parents would be more willing to practise family planning. It is no accident that the most successful family planning programmes are to be found in countries such as South Korea, Taiwan and China which first made progress in feeding their people and raising standards of living.

Can the same thing be achieved in the rest of the Third World? Can present miseries be relieved, and can mankind feed itself for all the foreseeable future? The answer is undoubtedly 'yes', but this answer is not as simple as was once thought.

People used to argue that if modern agricultural technology were fully applied to all the land in the world on which crops can be grown, a world population many times larger than the present one could be very adequately fed. They used to point out that there are many potential food sources which we have hardly begun to tap—the sea could be systematically farmed instead of being treated as a hunting ground; algae could be cultivated and processed to make human food; and leaf protein machines could extract protein from grass far more effectively than a grazing cow. High-yielding strains of wheat and rice could be produced (and such crops did indeed lead to a 'Green Revolution' in parts of Asia in the 1960s); and entirely new crops might have much to contribute, such as triticale (a wheat–rye hybrid) and the winged bean (a plant from south-east Asia).

All this technology arising from the work of plant breeders, chemical engineers and nutritionists can certainly make a contribution to world food production, but it is an illusion to think that, by itself, new technology can solve the problem of hunger— because without the people who must use it, technology cannot begin to be effective. The rather limited technological sensibility we have had in the past has tended to blind us to the restrictions and opportunities which this consideration places on us. Agricultural production is held back less by adequate technology than by the ways particular communities organise the use of land; whether they can arrange the selling of produce or its storage; and whether they can transport crops to market. Sometimes, farmers have to give such a large proportion of their profits or produce to a landlord, that they are discouraged from producing more. Sometimes, there is no way of selling surplus food, so there is no point in producing it. Sometimes the organisation necessary to provide fertilizer or set up an irrigation system is entirely lacking. Sometimes, people have values and ideals which cause them to see the problem quite differently from how we do, and there is no common basis for adopting a 'technological solution'.

Very frequently, new technology can contribute nothing until problems like these are sorted out—and even then, a technique imported from Europe or America without modification may not fit in with the way people choose to organise themselves. Finding or devising techniques that do fit the local pattern is an approach which has come to be referred to as *appropriate technology*. So, for example, it is often more appropriate to design farm implements which can be hauled by local oxen and which suit local soils and cultivation techniques than to import a tractor and plough from Europe.

All the possible new techniques mentioned above will need to be developed in variant forms appropriate to different people in different circumstances. A new food crop, however productive, will not be grown if it cannot be sold, or if there is no recipe for using it in food which people like or are used to. New kinds of equipment, like the machine for extracting leaf protein from grass which is currently being tried out in south India, will need to be carefully designed for village use if they are ever to be widely adopted.

This concept of *appropriateness* represents a step forward in our technological sensibility—a recognition that technology can achieve nothing without people, and that therefore technology must be adapted to how people actually live, and how they want to live. The ultimate in this is technology which so exactly meets people's needs and captures their enthusiasm that they achieve through its use far more than the originator or designer of the technique could have anticipated—people can, as it were, transcend the limitations of technology. Part of China's recent achievement has resulted from that, and other examples which come to mind are a village water-supply project in Lesotho and a brick-making and house-building project in Brazil.

Despite these successes, we should not look on appropriate technology as being by itself the solution to the Third World's problems. In many countries, people are hungry because political and economic arrangements prevent the available food from getting to them. In some countries, especially in Latin America, landlords find it more profitable not to cultivate all their land, or to grow crops like cotton or tobacco which people cannot eat. The solution to problems like these must be found through political change, both within individual Third World countries, and in the relationship between Western countries (where people often eat too much) and the Third World. Even the most appropriate technology can contribute little in circumstances where it is major economic and structural inequalities which hold back agriculture or imprison the poor in their cycle of misery.

Sources of Teaching Material

Help with speakers or visual aids dealing with the Third World can sometimes be obtained from local offices of Christian Aid or Oxfam (see your telephone directory). Oxfam has an Education Department (274 Banbury Road, Oxford, OX2 7DZ) which works with teachers and can provide project packs for appropriate technology activities in schools. Other ideas on this can be found from time to time in the magazine *School Technology*, or by contacting the Council for Education in World Citizenship (93 Albert Embankment, London S.E.1).

A wide range of teaching material on the Third World is available from Third World Publications (138 Stratford Road, Sparkbrook, Birmingham B11 1A11), and literature on appropriate technology is available from Intermediate Technology Publications (9 King Street, Covent Garden, London, W.C.2). Both these bookshops will mail a catalogue on request.

INDUSTRIAL DEVELOPMENT IN THE THIRD WORLD

STUART TAYLOR

Director, W. S. H. Taylor Engineering Developments Ltd.

and

GRAHAM EDWARDS

Chairman and formerly of Taylor Engineering Developments Ltd. and G. A. B. Edwards & Associates (Consultants), Director of Production Research Unit, University of Bradford

What new factors are involved in appropriate industrialization? Why are developing countries unable to produce their own vehicles and what kind of services would be needed to enable them to do so?

This kind of questioning led to the Trantor Project which includes not only the design of a new product but also the design of a new kind of factory unit based on cellular production, minimum investment of capital, durability and simplified manufacturing processes with low skill requirements.

Trantors are general purpose vehicles: they plough, take the full range of farming equipment, pull army equipment, carry 7 to 10 passengers and trail loads up to 10 tons at speeds up to 80 kph. They span three markets; those of the agricultural tractor, the Land Rover (or general purpose vehicle) and the small commercial lorry. Owners need only one Trantor instead of any combination of car, lorry, Land Rover and tractor.

The fundamental specification on which the design of vehicles to meet the particular needs of different local conditions is based, was drawn up following a two-year research study by W. S. Taylor of transport requirements in agriculture, forestry and in general tasks in rural areas. The technical specification based on the findings of this market survey was expressed in the following terms:

(i) There was a need for a vehicle which considerably increases the rate of work when carrying out those 'transport tasks' currently performed by the all-purpose agricultural tractor. The loads of tractors doing transport work vary from 'the driver and a few tools' to about 2 tons dead weight on the hitch, when using unbalanced trailers, and the drag from 10 ton trailer loads. Transport speeds could be from under 2 kph on slippery and tough ground to about 80 kph running light on good roads. The technical facilities required to accomplish this work were found to be, some form of pick-up hitch, some form of suspension capable of handling the very wide range of loads and speeds, hydraulic facilities to handle tipping trailers, a load platform on the vehicle itself and seating for one or two passengers.

(ii) There was a need for the vehicle to handle some standard agricultural implements to maintain its use when there was no transport work to be done. The technical facilities required to accomplish these tasks were, a simple form of three-point linkage with position control which would not suffer from any suspension sag, a standard agricultural power take-off (p.t.o.) which should preferably be 'live', there should be a high degree of manoeuvrability and good rearward visibility.

(iii) There was a requirement for a safety-cab with low noise level and driver comfort. The technical facilities required were, a structurally reinforced cab, some form of cab suspension and good seat and driver-control ergonomics.

(iv) The price of the product should be within the range of existing, similarly powered, all-purpose tractors.

The broad outline of the vehicle specification was summarised as follows:

(a) Cost factors ruled out four-wheel drive.
(b) The requirement for manoeuvrability emphasised the need for two-wheel drive with a short wheel-base and smaller front wheels.
(c) Good rearward visibility, watch any implements and to pick up trailers, the need to have access to the engine, the amount of gear-shift linkages, and the need to work with a light front-loader all emphasised that a more central driving position, rather than a forward cab, was preferable. This solution was a compromise because it had the effect of cutting down the available space for the load platform which was a further requirement.
(d) The need to work in slippery conditions, towing trailers with only two-wheel drive, emphasised the importance of the capability to handle unbalanced trailers. The rear hitch position had to be as near as possible under the rear axle to maintain longitudinal stability and, in addition, some form of weight-transferring device would be required to handle balanced trailers. A pick-up hitch and hydraulics for tipping trailers would also be essential requirements.
(e) The need for higher speeds was a critical feature if success was to be achieved. No system existed which gave the required suspension characteristics. A new system would have to be created if the objective was to be achieved. This part of the process was embraced within the technology design stage. Other aspects connected to the need for high speeds were a more efficient braking system and better lateral stability. These factors emphasised the need for as wide a track as possible and a reduction in rear wheel diameter but not tyre-width as well as the requirement to lower the centre of gravity, whilst maintaining reasonable ground clearance.

Design is difficult to catalogue; in this case it seemed to have four foci. First the suspension system and its hydro-mechanical principle had to be invented. Once it was known how the process could work there were, of course, several methods by which it could be transformed into mechanical hardware. Outside influences centred on simplicity in design and the minimisation of capital cost for equipment so that a big change in policy resulted. Considerations moved from a vehicle with four individual wheel units to only one unit concentrating on the hitch. Next, the suspension unit was rationalised with regard to stress calculations and designs for one-off prototype manufacture were drawn up. These turned out to be too crude in some aspects but the main object of the first prototype unit was to confirm that the system would work without having to invest in further costly items such as casting patterns. Finally, alterations were made to the prototype design as a result of obvious defects shown up in trials where factors such as wear and accessibility for maintenance were considered. The assembly design stage was largely one of selecting bolt-on proven units, devising methods of holding them together and establishing which units needed to be specially designed to produce the required product characteristics.

A feature of the Trantor Project is to offer manufacturing and assembly facilities for Trantors. The whole project is geared to the idea of assisting a developing country in widening and extending its skill base by offering simple fitting skills and machining skills. There are no components which require, for their economic justification, high-quantity machines processes, e.g. pressings, forgings, etc.

The second main feature of the Trantor Project is the design of a new kind of factory based on group or cellular production. The objective was for the factory to make the entire product and for manufacturing workers to join with assembly workers and administration in a manner which would encourage and develop identification with the market, quality and financial viability of the firm. Where opportunities for complete automation arose, there were to be no half measures and men were to be removed entirely. Wherever possible people were to be involved in skilful tasks, encouraged progressively to increase their skill and to become capable of working on a wide variety of work tasks. Stock in the system was to be either raw materials, finished components or finished products, and the system of production would be designed jointly from the technical and social systems viewpoints.

Assembly flow lines were therefore rejected: parts were produced with simple, inexpensive capital equipment in the form of easily available simple machine tools and specially designed jigs and tools. Groups of machine tools with associated jigs, tools and fixtures were set up to deal with families of components requiring similar machining, thereby reducing the time taken to change from one family to another and facilitating the standardisation of common features such as hole diameters, tolerances and material.

When designing the production system for the Trantor, it was clear that it would have to be produced economically in small quantities and that each Trantor factory would need to prove its economic viability where the number of employees would be about 100 to 150. At such a size the output of each plant is about 1000 units per year and if the market demand in a particular country, or region, becomes greater than this number it needs to be sufficiently viable in its usage of capital and give sufficient returns for it to be possible to duplicate the plant in another region of the country.

The Trantor Project is geared to the idea of assisting a developing country to operate a motor industry of its own which will satisfy the home market and provide export potential, and is based partly, wholly or firstly on Trantors. The service offered includes guidance in the conduct of comprehensive feasibility studies, in the design and assembly of Trantor products, in the design of work organisation and factory to embrace the precise needs, policies, facilities and abilities of the country concerned, together with a tailor-made training school to meet the needs of each stage of manufacturing, servicing and repair.

PART III

Viewpoints on the Curriculum

Introduction

This section relates themes explored in earlier parts of the book to the school curriculum and to several approaches to learning. Innumerable questions are posed: is there a need for fundamental change, for teaching which speculates about the future and illuminates the realities of life, sustains natural creativity and provides opportunities to get to the heart of problems? What is meant by 'problem-solving' and 'design', and how is inventive thinking stimulated and developed? How can technology be made more responsive to the wishes of ordinary people and how might the social responsibility of innovators be stirred? What factors inhibit the technological sensibility of girls? What is meant by education in technological literacy and how might it be developed?

Some polytechnic and university teachers respond to the challenge of change by proposing radical solutions: they plead for reorientation of effort to educate for survival, for conscious technological futures, and for wise stewardship of the earth's resources. They develop ideas and programmes for nourishing the inventive capacities of students (including children), for role-playing and decision-making in simulated situations, and for encouraging a more positive stance among teachers towards the technological society in which they live. Others probe the factors affecting the technological sensibility of girls and seek new ways of freeing imagination in the design of artefacts and systems.

School teachers are no less responsive, as the case-histories described in this section illustrate. Nor are they content with empirical approaches: their hard thinking about the nature of problem-solving, design, ways and means of developing new skills and attitudes in both girls and boys to ensure their material and spiritual survival, and their imaginative use of resources provided by new technologies, are hard evidence of what might be achieved, but not without adequate strategies for investment and development, and it is on this note that Part III is concluded.

Academic Pointers

EDUCATION FOR A TECHNOLOGICAL EXPLOSION

ADAM CURLE

Professor of Peace Studies, University of Bradford

Every 5 or 6 years the number of research findings in the natural sciences doubles the whole volume of findings previously recorded. If I compare my boyhood in rural England in the 1920s with descriptions of country life written in the early 1700s, and then contrast it with what exists today, I realize that more has altered in 50 years than in the preceding 200. I grew up without running water, gas, electricity, a radio, central heating (of course), a car (motors naturally existed though the roads were delightfully empty), but we were not particularly poor. It is not really necessary to stress the facts of change but we sometimes need to remind ourselves that the post-World War II world of television, antibiotics, nuclear fission, electronics, plastics, and jet travel is light years away from the period between the wars, technologically and in terms of the economy, and hence the society created by the technology.

The first implication for us is that we must be constantly adapting ourselves to a world which is forever on the move. This is very different from what obtains in the world's simpler peasant societies, though many of these too have been affected by the technological surge. There the wise man is he who best knows the lore of the past, who can most easily recall the details of ritual and tradition, for whom innovation is sacrilege. There are, of course, people like this in our own society, but they are a bewildered and ridiculed minority.

It is certainly confusing to live in a world of social change, for not only do the technologists—and hence the techniques of living—alter kaleidoscopically, but also the expectations and moral values which go with them. More than a generation separates us from our children: in a sense they belong to another civilization, but unless we can learn its language and its custom they will be lost to us.

I well recall the turmoil and excitement of the U.S.A. in the mid-sixties when many realized that they were training people for occupations which were obsolete, and strove to adapt their teaching to the needs of something they had never experienced, and indeed could hardly imagine. The implications of all this for us are enormous. How can we in Britain impart the knowledge, attitudes and skills to do jobs which no one has even thought of? How can we prevent our preoccupation with the past, which is really all we have to offer, from being a shackle holding back our children from entering the future?

It may be objected that education should not be concerned with preparation for work but in the wider sense for life, leading out human potentialities. I really mean it in both senses, but unfortunately no one can ignore the fact that educational systems everywhere are largely geared to producing the types of people society feels to be needed to carry out a variety of functions, and that an educational qualification is to a considerable extent a job qualification, in some cases, e.g. medicine, completely so.

The contemporary paradox is that while obviously no education can be carried out without imparting knowledge, it must be given, and hence acquired, in such a way that it is thought of as malleable rather than firmly established, expendable rather than sacrosanct. I was taught physical anthropology largely on the basis of theories derived from a study of Piltdown man—who turned out to be a forgery! Had I remained in that field I would have had to unlearn and then relearn almost everything. I recall also that a whole generation of physics professors at Harvard, where I taught for some years, were eased out of their jobs because they refused to recognise that Einstein's model of the universe had replaced that of Newton. But, provided we are ready to discard or modify it, all knowledge is useful in focusing the mind on various problems and emphasising method and rigour in study and research.

Are there many teachers, however, who are prepared to say that what they are teaching may be wrong? Most, I fear, are far too much identified with their knowledge—they owe to it their positions and doubt would spell insecurity.

How, then, does one teach openness, receptivity, flexibility and imagination? These are the qualities which are supremely important for facing an unknown future. I should perhaps add stability, not in the sense of standing firm but of remaining unharmed by what has been termed 'future shock'.

We have adjusted very badly to the challenge of change. And a glance at the educational system shows, at least to some extent, why. Until very recently we have been the most under-educated of all the industrially advanced nations. What education has imparted, moreover, is constrained by a narrow and archaic examination system which channels and divides human interests and abilities in a way which is both against common sense and against adaptation to the infinite complexities of our changing world. Our educational administrators, as a body, have failed to break the Oxbridge spell: we are still churning out an elite of little humanistic scholars. An exaggeration of course, but one which is justified if we compare educational developments here with those in America, West Germany, Sweden, or Japan: our specialisations are geared to a limited range of largely obsolete forms of employment.

Let me conclude by being more specific. What we need, I believe, is an education which informs us and our children about the world we live in, encompassing all its major dimensions—scientific, historical, social, economic and psychological. The teaching, where appropriate, should demonstrate the curve of events from the past to the present, and thence by speculative extrapolation into the future. There should be music, poetry and philosophy, to sustain the natural creativity which is normally dulled by education. There should be the chance to specialize for those with special interests, for that is how one gets to the heart of problems. But *not*, repeat *not*, to pass examinations.

But what, it may be asked, does this do to prepare young people for work? It does all that matters, for the ultimate necessity is a mind which is open and alive. Given these vital qualities, specific skills can easily be learned, but without them nothing worthwhile can be achieved.

TECHNOLOGY AND TEACHERS

GEORGE JACKSON

Professor of Mechanical Engineering, Brunel University

'Now you see what will happen to you if you fail your exams.' Teacher's comment at the end of the works tour struck home to his sixth-form class. So that was what happened to the unqualified, to those who had been relegated to metalwork and technical drawing, to the engineers who had not been good enough for mathematics or science courses. There they all were, in low-status jobs and doing humdrum work in an unwholesome environment. Neither challenge nor opportunity in the work, those few with a remaining spark sought fulfilment in outwitting the system. There but for the grace of the examiners would they go too.

For the school party's host and guide, the firm's training officer, the teacher's remark put an end to his recruitment hopes. Coming from a fellow educationist, it hurt. Industry plays, and for a very long time has played, an important part in the educational system of the country. The training officer's job is to develop the skills and talents of his firm's employees, not only skills in making things and doing things but also talents as critics and as originators and communicators of ideas. The object is of course to ensure the firm's survival but employers and training officers alike have no doubt that personal development is synonymous with the public good. The distinction between education and training, clear enough to some of the donors, is lost on the recipients. If they see a distinction it is that they are paid for one but not for the other. Educators, inside the teaching profession and elsewhere, are separated by the barrier of a shared interest.

If the satanic-mills—big hands and tiny head—view of technology is correct, the time is long overdue for reviving technology with an injection of talent. If it is false, the detractors are as immune as the rest of us to persuasion by fact or by logic. It is beyond question that there are pupils, many and talented pupils, whose wish is to have nothing to do with technological activities. Whilst it is true that teachers are far from being the sole transmitters of opinion, it is unavoidable that teachers of technology-linked subjects at all levels play a part, by example and precept, in shaping attitudes. School teachers of woodwork, metalwork, technical drawing, engineering, design and technology; work-experience supervisors, industry-based teachers of craft and technician apprentices, industrial tutors of sandwich course students; college teachers of day-release, diploma and certificate students and university technology teachers are obviously concerned. Teachers of art, craft, design and science at all levels cannot avoid being involved. Perhaps most important, if anything is to be changed in this diverse group of autonomous individuals, are those responsible for the selection, admission and preparation of the next generation of teachers.

A minority of these teachers, across the spectrum of academic subjects and student age groups, have convinced themselves from their own experience that a new integration between theory and practice improves their teaching. For most of us it is easier to carry on doing what we have always done in the way in which we have always done it and it is unnerving if a teacher from a contiguous subject area seems to be making a take-over bid. It is of the nature of technological activity that it calls on a wide range of knowledge and skills and offers scope for a wide range of abilities to be applied by a team of collaborators. In the adult world it takes all sorts to plan, design, construct and operate successfully a motorway or a fish-finger plant. Involvement in a project in the student world is as near as most people come to seeing in microcosm the totality of an enterprise. The educational value of project work has been widely recognised for at least 20 years. A specialist teacher attempting to put a subject in a technological context needs support, but colleagues are suspicious, headmasters are conservative, timetables are restrictive, entrance requirements for the next educational stage are inflexible, and the traditions of teacher training are not easily changed. The educational system is incestuous.

The contemporary educational argument about the proper place of technologically oriented teaching and about the best methods of preparing children for coming to terms with technology can probably be traced to the same impetus which gave rise to the 1951 exhibition, 'The Festival of Britain'. A quarter of a century of debate and experiment has given birth to technology courses for teachers, sandwich courses, engineering examinations at A-level, CSE technology, industry training boards, design-make-and-test projects, science fairs, technological universities, interdisciplinary studies and so on, amounting in total to a generous over-provision of resources in relation to the student demand. The argument remains unresolved. Few are prepared to admit that a liberal education can be given in a technological context. New curricula, new teaching methods and new examinations devised by forward-looking, experienced and capable teachers succumb to the institutionalisation of learning when the originating enthusiasts leave the scene. Their efforts deserve better support and fewer obstacles need to be put in their path.

TECHNOLOGY APPRECIATION IN TEACHER EDUCATION

MICHAEL SAYER

Lecturer in Technology Education, University of Keele

How many of us really appreciate Technology—that is, Technology in its broadest sense? One could easily be trapped into thinking, for example, that 'Educational Technology' is restricted to the gadgets and mechanisms marketed by the manufacturers of these articles, advertised in glossy handouts at exhibitions; but, as any efficient teacher will know, the technology of education is wider and deeper than this. It includes things like efficient use of 35 minutes to give a comprehensible lesson, efficient cleaning of the blackboard using least time and effort, and efficient ways of assessing the understanding of what has been taught.

Technology is efficient use of resources, and good Technology is the discipline of designing things for the benefit of people—things not beneficial are bad Technology or, at worst, misapplied science. It is important to appreciate the social interaction implicit in Technology: Technology (and the processes of design and of utilizing natural resources) cannot exist without people, but in an unpopulated world the laws of science would still apply, although undiscovered and unexplained.

The view of Technology as merely the application of mechanisms to social groups is shown to be full of hazards: Sir Alan Cottrell, writing in *Physics Bulletin* of July 1975 (page 303), says:

> Purely technological attempts to solve social problems, all too often produce not the effects hoped for by their designers, but quite different ones grasped opportunistically by society. This is no new insight, but we are today I think still too dazzled by physical science and technology to recognise it sufficiently. Yet examples of it can be traced back a long way. The use of the miners lamp at first increased casualties in the mines, because it enabled miners to work deeper and more dangerous seams; and the introduction of anaesthetics at first increased hospital deaths because it encouraged surgeons to try more difficult operations.

Since Society as we have come to know it is dominated by Technology good and bad, it has become urgent for young adults to understand the processes of Technology, the disciplines of design and optimization. One way of promoting this vital understanding is through case studies of particular designs, large and small scale, historical and contemporary. In this way we can hope to form a balanced appreciation of the processes of Technology and a rational judgement of the social effects. E. G. Mesthene, in *Technological Change* (Mentor Books, 1970), contrasts two extreme views:

> taken together, the two views may be seen as the latest stage of the eternal battle between God and the Devil. In the first view, technology is invested with an omnipotence heretofore reserved to the Almighty. In the second view, technology emerges as the modern counterpart of the Devil, responsible, as the Devil has traditionally been, for man's eternal inhumanity to man.

Technology appreciation, if educationally effective, ought to lead us to Mesthene's third view:

> technology as such is not worthy of special notice ...: it has moreover been recognised as a factor in social change at least since the Industrial Revolution; it is unlikely that the social effects of modern technology—even of computers, for example—will be nearly so traumatic as the introduction of the factory system into 18th-century England.

Taken together, the preceding paragraphs can be read as a case for educational studies in History of Technology; it is an area of study equally attractive to both Arts and Science—biased pupils and, for the Arts people, one can introduce the basics of Science disciplines by technical investigation and simple calculations, whilst the scientific pupils can learn something of the way designers actually design things—itself a social process in method and effect.

Technology is, in fact, an identifiable discipline at the interface of Arts and Sciences, drawing from the processes of both and linking the two in social and technical interaction. G. L. Glegg, in *The Design of Design* (Cambridge University Press, 1969), defines three processes of thought in the mind of the technological designer:

1. The Inventive—the sudden unsought inspiration that occurs in the subconscious imaginative mind.
2. The Artistic—the sense of fitness or suitability, evolved through experience, that can develop the inventive idea.
3. The Rational, 'holding the power of veto over them all'.

Technology Appreciation, then, is not a product of purely technical study; it requires an overall view of the development of several technologies in a social and historical context, together with a quantified depth study of at least one of them. It requires also an appreciation of both rational and intuitive thought-processes, preferably with some directed experience of artistic and scientific methods; and it requires understanding of industrial processes in the context of both contemporary and past society.

Like music appreciation, Technology Appreciation is open to anyone for personal enrichment but, like music appreciation, the enrichment is heightened through the experience of participation and some technical study.

SCIENCE IN ACTION

DERYK KELLY

Director of the Engineering Science Development Unit at the Department of Education, University of Loughborough

> Applied science must form a protective wall around pure science. There are many people who deny that there is a boundary between pure and applied science. I do not agree; there *is* such a thing as pure science. Its sole aim is the extension of knowledge. In our society it can exist only if it is protected by a healthy applied science.*

* Professor Dennis Gabor, *Esso Magazine*, vol. XXI, no. 4 (Autumn 1972), p. 25.

Despite the intensive curriculum development efforts of the last 20 years, and despite the comparatively large sums of money expended by the Nuffield Foundation and by the Schools Council, there is little evidence to show that there has been any significant increase in interest in science among secondary-school students. The spate of articles in Physics Education over the last 12 months bears eloquent witness to the concern which this failure has aroused among science teachers in both schools and universities. The nature of the concern and the remedies proposed vary widely, but all are linked by a common fear that the days of science teaching in our schools might be numbered.

The Objectives of Science Teaching

The educational world in general is now paying much more attention to aims and objectives than was the case in the years immediately following the Second World War; unfortunately, in the case of science teaching, such concern has not gone very deep. While it is true that most of the syllabuses and schemes of work which have been produced in the present decade have formally listed a set of educational objectives, most of them have started from the premise that science teaching (or learning) is a good thing in itself and needs no justification. Students, however, are not convinced and appear to be voting with their feet against such an *a priori* assumption.

When the purposes of science teaching are discussed the principal argument appears to be that science is an important element in our culture and that, therefore, children should be made aware of its methods and its designs. The practitioner of any discipline, be it in the arts, humanities or the sciences, is naturally enthused by the coherence of that discipline and by the elegance of methods used by its devotees. Such an appreciation demands a high degree of sophistication and a high level of competence in the discipline; almost by definition such high levels are rarely attained in school studies. (One might even question the extent to which the bulk of university or polytechnic students achieve such desirable states.)

An important corollary of this lack of expertise is that the historical development of the discipline arouses even less enthusiasm among children; moreover, although the work of scientists has changed our view of the world and of man's place in it, it is sadly true that few children can grapple with such complex abstractions. Although older science students, particularly sixth formers, are often very much interested in the history and philosophy of science and in its social and moral import, science is usually rejected long before this stage is reached. There is thus evident dichotomy of interest between most young people on the one hand, and teachers and curriculum developers on the other.

Engineering Practice or Science in Action

What are industrial scientists but people who apply scientific knowledge and expertise to the solution of practical problems; they may be described as applied scientists or equally well as engineers or technologists. On the other hand, engineers who migrate to universities to undertake research might well be more aptly described as scientists. It would, therefore, appear that to a significant degree we are bedevilled by tradition and by name.

Whenever engineers or scientists tackle industrial problems they use not only science but experience, intuition, and a vast range of economic and social resources. In place of the comparatively neat, tidy, controlled situations in which they are seeking unique answers to fairly clear problems, they find themselves faced with problems to which there are innumerable solutions, which are to some extent at least determined by factors completely outside their control; what is worse, they rarely have as much information as they would like and are often forced back on previous experience and downright intuition. In other words, they are facing a human situation rather than an ideal situation.

Curricular Consequences

It is difficult to over-emphasise the implication of this point for education. Whatever the future careers or education of our sixth-form students, they will be faced in both their working and private lives with the necessity of solving practical problems, having limited resources at their disposal and with a woeful lack of information to guide them. To some degree they will be called upon to act like engineers. If this generalisation has any real significance, then the schools have a responsibility to develop those abilities in their students which will help them tackle such problems.

The consequences for science teaching are fairly obvious. If we couple the desirability of imparting to the youngsters some appreciation of the way in which society satisfies its material needs (i.e. give them some understanding of science and engineering as they are practised), with the necessity of developing problem-solving abilities, then our science courses must move away from the pure and the academic towards the applied and the relevant.

The question which now poses itself is whether it is possible to achieve these aims within the context of subjects labelled physics, chemistry and biology. We may, of course, define these terms in any way which is suitable and relevant, but they are so strongly bound by their traditional connotations of isolated academic study that it is doubtful whether any such attempt could be even marginally successful. The time may well have arrived for them to disappear from the school curriculum. They lack even the justification that the traditional craft subjects possess in that the latter do at least develop significant practical skills in their own right.

In place of physics, chemistry and biology we need a set of studies in the shaping of man's physical and material environment, studies to which physics, chemistry, and biology teachers can each make a significant contribution in co-operation with colleagues from other faculties.

DESIGN EDUCATION

KEN BAYNES

Research Fellow, Department of Design Research, Royal College of Art

There has always been a large area of common ground between those who have promoted the study of technology in schools and those who have spoken of 'design education' as being an essential core subject in any secondary curriculum. By now, it may only be semantic problems which really divide them although, in practice, there are differences of emphasis. Rather than argue about this, I would prefer to use the present essay to explore a topic which I believe to be of central importance for every teacher concerned with either design or technology; one which sometimes divides both groups.

Nearly all arguments for design education point out its social utility. It is generally seen as essential in an increasingly industrialised world which depends to an ever greater degree on technology. But not everyone interprets its importance in the same manner. At the most simplistic, the approach is to say that engineers, architects and designers depend on an 'informed public' to be able to work effectively. Frequently, this claim is allied with another: that our present economic plight is due to the ignorance and apathy shown by the British people towards design and technology. In this picture, the aim is to reform the public. It is an old argument which has reappeared with regularity ever since the Industrial Revolution. However, there is another way of looking at the problem. It is to say that engineers, architects and designers have not succeeded in creating a world which is responsive to the needs of ordinary people. In this alternative picture it is also important that the public should know about design and technology but for a different reason. The purpose is to give them the knowledge and skill to control and direct the experts who should serve them.

We are thus faced with two completely different views of society. Related to it are two completely different interpretations of the role of technology in the modern world. The dramatic thing for the teacher is that the view he adopts will necessarily affect his approach to education at every level. It will affect detailed teaching methods and the kind of work done at the workbench as much as it will affect such things as curriculum balance and the significance of design and technology as an aspect of general education.

How then should we approach technology?

My own belief is that any approach to technology which reinforces its 'magic' as against its fallibility is counter-productive. Technologists often make mistakes. Designers often make mistakes. People need to be equipped to understand this and to respond in a way which is creative as well as critical. It brings us back to such ancient educational ideals as the fostering of independence and responsibility. It is here that I believe the design education 'movement' has a special contribution to make to the understanding of technology. What is emerging is the idea that the primary responsibility of the design teacher is to encourage the development of a rounded personality. Specific content is a secondary factor.

In 1973, there was an important conference at Horncastle organised by the Department of Education and Science. One of the speakers was Philip Roberts, then the schoolteacher secretary of the National Association for Design Education (NADE). In his contribution he attempted to sum up the present scope of design education in schools and to look to the future. Here is something of what he had to say:

> A natural result of looking at social needs and design problems (instead of preconceived solutions) is to make design studies educationally more valid, and socially aware and responsive. In some cases there has been a move towards a greater consideration of *systems*, with discussion of the relation between parts of the system, and their social implications. For instance, instead of starting with the idea of making 'a car', the start is made with the idea of a transport system into which components must fit. This naturally brings into focus such issues as pollution, the depletion of resources and the value of the individual and society in relation to technology, economics, fashion and so on. This kind of analysis, which really should be the essence of design education, could alter radically the nature of education in relation to society: in its implications, this genuinely *is* a revolutionary approach.
>
> The next step in design education is obvious and reasonable but it has not been defined by secondary teachers in practical and explicit teaching terms. What is needed in the future is design education at the level of the whole community leading to the participation by one and all in environmental decisions. In the majority of cases a social and political structure for design education in this broad sense has not been considered, though it is hinted at in some syllabuses which are based on man and his environment.

At a subsequent Conference*, this time organised by NADE, I chaired a small working group of teachers and lecturers whose task it was to examine the nature of the interrelationships between 'Thinking', 'Learning' and 'Design'.

Amongst a number of points made in an introductory paper were the following:

* Woolley Hall, January 1976. The papers from this Conference are to be published by the National Association for Design Education.

1. We are not attempting, in school, to produce potential entrants into the design professions. Our interest in design activity is as a learning medium—a formative experience for the individuals who undertake it. This means that we may take from the adult world of design those things which we see as educationally useful, add to them or subtract from them. It is important to clarify this relationship because, at the moment it is a serious source of confusion to pupils, teachers and professional designers.
2. In a democratic society we are seeking to encourage a type of individuality which is distinctive and creative, but at the same time, responsive to the community. This means that we should engage children in activities where they are:
   (a) Free to think for themselves and to act on their ideas. That is to experience the results of their own decision-making. This provides a real learning situation.
   (b) Involved in tackling the real and pressing qualitative problems of life.
   (c) Able to see how their ideas relate to and possibly conflict with the beliefs and requirements of others.
3. Design activity is essentially about 'travelling into the unknown', about making decisions which will determine the future quality of life. In education, this adventure must be genuine if it is to promote learning. Neither the teacher nor the pupil may know the answer in advance. This is a particularly valuable aspect of design education.
4. There are no '*right*' answers in design. This challenges the role of the teacher. He becomes more the pupil's tutor or consultant, providing information where he can, joining in the search where he can't. Potentially this is the most dramatic advantage which design activity has to offer both the pupil and the teacher who then join together to learn together.

If these ideals are accepted, then it is up to the individual teacher to judge how far his own approaches are appropriate.

What is striking is the degree to which teachers challenge any narrow concept of design or technology teaching; they provide no place for the kind of 'experimental' project work in which children undertake small pieces of component design without appreciating the larger world of social, aesthetic and ethical considerations. The urgent need is for us all to experience the inevitable interaction between technology and the qualitative aspects of life. It is this alone which can humanise and dramatise its role in modern civilisation and culture. Particularly for general education, it is this which must, in my view, be the crucial issue for teachers.

ENGINEERING DESIGN

JOSEPH BLACK

Professor, School of Engineering, University of Bath

Most people fail to distinguish between 'science' and 'engineering': man landing on the moon was described as a 'scientific' triumph, but if the rockets did not lift off it was an 'engineering' failure. The scientist is primarily concerned with discovering, codifying and understanding the nature of the physical world. He seeks to solve problems which have definite answers, and his output is new knowledge in published papers. The goal of the engineer is to utilise knowledge of the physical world for social benefit: the problems he seeks to solve have many possible answers, from which he must attempt to select an optimum solution, which will end up as 'hardware', i.e. a product, device, process or system.

In the early days of technology, and reaching a peak with the great nineteenth-century engineers, one man could cope with the whole process from identifying the need right through to selling the manufactured product—Watt with his steam engine, Brunel with his railways, bridges and ships. But modern technology is so complex that whole teams of 'experts' are involved; we have thermodynamicists, stressmen, materials men, vibration men, test men, value analysts, environmental engineers, ergonomists—to name but a few. Then we have for each component of any overall product the specialist supplier of assemblies, components and detailed parts, with similar subdivisions of experts. The men who can bring all these aspects together, or 'synthesise' them, are the engineering designers.

Thus 'engineering design' can be described as the process of identifying a need, then applying all the scientific, and engineering principles and techniques, invention and flair, for the purpose of defining a device, a process or system, in sufficient detail to permit its physical realisation, which will satisfy the need. The data available to the designer will almost always be incomplete, inadequate, irrelevant and contradictory; there will be severe constraints of cost and time. But therein lies the attraction and challenge of design, since it demands personal judgement, and remains an art, even though so much of the background, education, training and problem-solving may involve considerable science, technology and mathematics. A capacity for human relationships is also a major requirement arising from dealing with the 'customers' whose needs must be met, and the co-ordination, control and resolution of conflicts with the design team members.

Engineering Design is often confused with 'engineering drawing', which is only one of its languages. This fallacy is comparable to thinking of the composing of music as merely putting little ink blobs on five parallel lines. Design, like composing, is largely a mental process, and the drawings at this stage are, like his notes, the recordings by the designer of the decisions being made. They are a pictorial extension of what he is thinking, and for physical space relationships they are still the best method, often as free-hand sketches—but functional relationships may be better expressed in symbols, mathematical and computer, or flow charts, circuit diagrams and block diagrams. Engineering drawings become much more important at the detail design stage, and are the communication for manufacturing instructions (though now we can produce some three-dimensional components by direct instruction from computer to numerically controlled machine tools, with no drawings involved).

Briefly, the designer's first task is to take the need, as expressed by the 'customer' (who may be society as a whole, a section, or an individual), or in the case of a new idea, the potential use, and convert this need or use into an engineering specification, which can be assessed on the basis of resources required to produce the device, the likely cost, and the time to achieve it. (The subsequent stages of detail design for manufacture, though vitally important, do not concern us here.)

In our Final Year Design Course we simulate this process by making our students work in teams of ten to twelve, who elect a 'Chief Engineer', and are set tasks for which they must produce a 'feasibility study' within a few weeks, on which an 'order' can be placed. Each team then carries out a complete design study for realisation with working drawings, costs, a full report, and models. These tasks would be equally suitable for any team of potential technologists, though with school children one could not expect the same level of mathematical and technical analysis. Some typical examples are given:

1. *Fertiliser distributor* for wide 'bout' widths: standard methods have a width of 10 m. There would be considerable economic advantages in widening this to 15 m while maintaining an accurate and controllable distribution.
   Solution: A novel pneumatic conveying system.
2. *Rescue device for deep-towed sonar scanner.* An expensive under-water sonar scanner could be lost if the towing cable breaks. The system should detect breakage and lift scanner to surface.
   Solution: Lithium hydride/water generates hydrogen which fills folding bags, automatically tripped by change in angle of towing bracket after breakage. (This won the First Prize of £500 in the Design Council (Molins) national competition for the students involved.)
3. *Removal of piles of old cars in breakers' yards.* These ugly 'mountains' must be reduced in size for economic transport to steelworks.
   Solution. A mobile car-crusher utilising a novel method of first 'crimping' the car body, thus reducing the crushing load. This crusher can visit the yards, in place of taking car bodies to crusher.
4. *Frozen-food transport system.* Present frozen-food transport is by large refrigerated lorries, causing traffic congestion, and severe loss of 'cold' when unloaders open and enter the cold chamber.
   Solution. 'Marketing' study determined optimum size of load, much smaller than for present lorries. New small heavily insulated van, with novel storage of food in plastic tubes, extracted from cab without necessity to open large doors. Stock prepared in pallets in 'cold room', thus reducing demand on refrigeration in van.

There are many more examples, but the common feature to all the tasks is based on the following aims and objectives:

1. To prepare each student to be a member of a design group in a firm or organisation.
2. To make the student sympathetic to the difficulties experienced by the design group if he should join other groups within the organisation.
3. To model a project to include conception, planning and progress, with time and cost as constraints.
4. To simulate the experience of working in a small design team.

After taking the course the student should be able to:

1. Convert needs to 'engineering requirements'.
2. Transform 'requirements' to 'job specifications'.
3. Incorporate team decisions into his particular designs.
4. Evaluate and analyse the function of elements of machines, products and processes.
5. Design elements for economic manufacture taking full account of essentials to good design such as ease of operation, maintenance, reliability, noise and aesthetic appeal.
6. Present concise reports covering the whole design process.
7. To 'sell' his design by models, display, brochures, etc.

The assessment of each student's efforts is then made as follows:
How well did the student

(1) convert 'needs' into specifications?,
(2) seek the background information, market requirements and existing solutions?,
(3) evolve possible systems to satisfy the specification?,
(4) present clearly proposed solutions to meet the specification?,
(5) justify the selected solution by evaluation of performance, cost and manufacturing methods?,
(6) carry out analysis and computation for the elements and components?,
(7) prepare descriptions, drawings, and models to convey his ideas, and to prepare for manufacture?,
(8) write a report and critical assessment covering his complete design evolution?

The enthusiastic response of, and the immense efforts put into their design work by our students confirms our belief that design is a challenge which is welcomed by the aspiring engineer and technologist, largely because it forces him to use all his knowledge, skill, flair and invention, and to seek out new information for himself. There is no right or wrong answer, so that the budding designer can, and often does, produce a better solution than the so-called expert—but experience does help, so it is never too early to start to design, and learn by one's mistakes.

ENVIRONMENTAL STUDIES

SIR FREDERICK WARNER*

Partner of Cremer and Warner, Consultant Engineers

Pollution over the years has had its critical events; the cholera epidemics of mid-nineteenth century London from sewage in the Thames; the 1952 great 'smog' in London; the poisoning of Minimato Bay in Japan; the radiation persisting after the explosion of atomic bombs. Each of these events had firstly publicity, public reaction and indignation, then pressure on legislation, enquiries, reports and regulations.

Between the crises, there is little of news value, and limited public interest. The environmental writers plug away about destruction of ozone in the stratosphere by supersonic planes and fluorocarbons or accumulation of zinc and cadmium in limpets or the newly discovered rare cancer from vinyl chloride monomer—even the old one from B-naphthylamine, whose manufacture ceased in this country 25 years ago. Nobody pays much attention and it is a pity. There is still a great deal to be done and learned about pollution but the research for the most part is unspectacular and tedious. What gleams of light there are shine only for the dedicated scientist or engineer.

Environmental studies at school or after have to find justification in educational terms. They do not stand on their own feet but on the shoulders of other disciplines. At whatever level they are taken, the appropriate basic studies are a prerequisite, whether the need is for qualification as a technician or as a full-blown scientist engineer or architect. With a good steady interest in chemistry, physics, biology or mathematics the superstructure can be more easily supported. As each new discipline has to be approached, fresh connections and understandings are established. Indeed, the first difficulties come when an attempt is made to put numbers to phenomena. The crude data of pollution at

* Sir Frederick Warner is a Visiting Professor in Environmental Engineering at University College, London and in Chemical Engineering at Imperial College. He is Pro-Chancellor of the Open University and consultant to the Oxford University Course on Environmental Control. He was recently adviser to the Court of Inquiry into the Flixborough Disaster and organized many of the technical investigations. He is a member of the Royal Commission on Environmental Pollution.

first appear to have no simple relations or derivations. They are complicated by the behaviour of the environment into which the pollutants are introduced.

The concepts are not easy. The distress caused by pollutants in air or water can vary from rapid death to chronic illness, although these can be avoided by moving away from the affected environment. The measurements are normally in concentrations and these are often given as limits. For example, toxicity is often given as LC_{50}, the concentration needed to give 50 per cent mortality after two days. The ability to absorb oxygen from water can be given as biochemical oxygen demand (BOD), which is expressed in parts per million so that it can be quickly compared with the amount in water saturated with oxygen at normal temperatures, about 10 ppm. But to make the comparison or to predict the effect of a polluting discharge, a mass balance has to be made, i.e. the capacity of the receiving environment has to be measured. This is normally done by water engineers who are trained in the methods available. They have been able to measure through almost the summer of 1975 the flow of fresh water in the Thames over Teddington Weir into the tidal estuary. It has seldom exceeded 9 m^3/s (or 9 tonnes per second), usually fully saturated with oxygen so that it carries roughly 9 tonnes/day (we make a lazy man's calculation that a day has about 10^5 seconds and oxygen solubility 10 mg/l). This 9 m^3/s has to take treated sewage totalling four times this flow and needing at a guess 250 tonnes of oxygen to meet its BOD. The fresh water is quite inadequate and it has been touch and go whether the contribution from the sea water coming in on the tides and solution out of the air could keep the river from losing all its oxygen. When this happens, the fish take avoiding action and keep away from the sections which will not allow them to breathe. The prediction of what will happen at any given time is made impossible by the inertia of the system—at times like these the water from Teddington can take 50 days to travel the 120 km to the open sea.

Humans do not have the same option to change what they breathe and so it is necessary that pollutants in air are reduced to concentrations which cause neither acute distress nor chronic bronchitis. I think it is fair to say that we do not have in the U.K. human health problems from air pollution. The one exception is the voluntary pollution from smoking which causes 50,000 deaths a year. The way industrial pollutants are diluted and dispersed demonstrates the mobility of the atmospheric environment. The concentrations which are measured depend on the period of sampling since the mixing varies as the wind changes speed and direction. It is also affected by topography and other meteorological conditions. To some extent, noise is diluted in a similar way as another mainly airborne pollutant.

I put the attraction of environmental studies into three areas—ethical, aesthetic and intellectual. There is a duty to respect the rest of creation and make the best use of the resources we have been given; there is the pleasure to be had from clean and comely environment: and there is no end to the stimulation which comes from new visions continuously revealed by greater understanding.

RETURN TO REALISM

JAMES GORDON

Professor of Materials Technology, Department of Engineering and Cybernetics, University of Reading

I dare say that a good many boys take up science partly in order to get away from those boring, overwhelming and unpredictable entities called 'people'. And then, science can represent an adventure of exploration into regions satisfactorily remote from the exacting everyday world. 'Science' is about astronomy and electricity and wireless and nuclear physics and the more alarming sorts of chemistry. This attitude to science can both attract and repel. It attracts intellectually in the same sort of way that polar exploration and small boat voyaging and space travel attract the physical escapist. Furthermore, in an age of weakening religious faith, this provides a comforting eschatology. Space-time may be a surrogate for the medieval heaven and, in the minds of many sixth-formers, Albert Einstein is no doubt a God-the-Father figure. Even more significantly, high science provides a body of Eleusinian mysteries, an arcane knowledge, an occasion for feeling superior. After all, we really do know more about the nature of the physical world than our mothers, or the history master, or even the Captain of Football; though we are decent enough not to rub it in. In our secret world, we have got our own back on 'people'.

Although this sort of knowledge may license us to mend, or break, the television set, its relevance to the rest of the life of common day is apt to be rather tenuous. Indeed and contrarywise, the mere fact that the science which is generally taught *is* remote from workaday life and human problems is just what is likely to repel the humanist. Its appeal to girls, in particular, is usually rather small: and, after all, women do make up half of humanity.

Naturally, a moderate dose of Galileo and Darwin, of relativity and electronics is good for humanists, stretching their minds and giving them a sense of proportion. It is part of the folk-lore of giant-killing—we have too few heroes nowadays! But! but! but! it is not whatever the science masters may say, an education in itself. In fact it is very much worse than the old traditional classical education. More even than the classics, it imparts a sense of superiority without actually communicating very much useful knowledge. The classics did, after all, combine that which is beautiful and intellectually taxing with the study of interesting people and important events.

Once upon a time, when the prime duty of science was to do battle with wolves in bishops' clothing, it behoved scientists to conduct themselves high and disposedly: to be concerned with dignified subjects like cosmology and the fundamental nature of matter. This was fair enough at the time, but now we have a new heaven and a new earth, complete with evolution and anaesthetics and birth-control. These battles are won, at least for the time being, and we might be wise to come down from the clouds, for we could be in danger of making asses of ourselves by being too high-falluting.

The greatest of electrical physicists, Michael Faraday, was also one of the greatest of teachers of elementary science. For the best and most loved of his Christmas lectures to children he chose for his subject, not some aspect of electricity and magnetism, but an ordinary candle. Living in a candle-lit age, the children who came to the Royal Institution must have been so familiar with candles that they had never stopped to think about them at all. It is a measure of Faraday's stature that he could not only discover electro-magnetic induction, but also make a common candle fascinating, both to the children of 1850 and also to us today. As an introduction to physical science these lectures are superb. Perhaps they should be given again, verbatim, and on television.

I suppose that every child who has been in the Science Fifth has done an experiment on Hooke's law and Young's modulus but even those who decide to read engineering have forgotten all about it by the time they get to the university. They know, with absolute certainty, that the subject is both trivial and boring. My own experience is that university entrants with A levels in Physics and Chemistry can be counted on to have some reasonable knowledge of heat, light, sound and electricity, in addition to a certain amount of chemistry and mathematics. But they arrive without even a tadpole-wriggle of an idea about the nature and behaviour of common solids. One begins to teach from a state of primeval darkness. If this is so with engineering students, what must be the case with the rest of the population, not to mention doctors and biologists?

Probably a good deal of harm, in the long run, has been done to the image of materials science by the idea that the subject is really only significant in high technology, in transistors and radar and fibre optics and nuclear reactors. These things may be well enough for those who are interested in that kind of thing but they cover only a tiny, and relatively unimportant, corner of the subject. Saucepans and shoes are *more important* than transistors or nuclear reactors and, to those with eyes to see, usually more interesting. Materials science is relevant to bridges and aeroplanes and racing cars, of course, but it is still more relevant to houses and furniture and carpets and curtains and agricultural machinery and packaging and roadmaking and surgical protheses.

Until quite recently our whole technology has been shaped and conditioned to fit in with the few materials which happened to be available and about which we really understood very little. We are now just beginning to understand what materials science is about, how to learn from nature and how to design materials to suit ourselves. This is taking the lid off Pandora's box with a vengeance. What will emerge is wholly uncertain but it will neither be dull nor insignificant. One can argue that all this presents the greatest opportunity for creative and imaginative technology which has occurred so far in history.

However, only a minority of the human race end up as creative scientists and engineers. One has to think about the broad educational value of materials science as a subject taught to the many. I believe that the chief objection to 'science' in general as an educational subject is its tendency to spawn specialisms. From quite an early stage, different kinds of scientists have few ideas in common and they are barely able to communicate with each other, let alone with the laity. Essentially, materials science may be regarded as a unifying discipline. The same ideas about materials can run right through engineering and biology and medicine and astronomy and textiles and mining and geology and architecture and archaeology and agriculture and ordinary house-keeping. All these activities—and a great many other ones as well—depend upon the properties of living or dead materials. Materials science is essentially the catholic, the down-to-earth, the unifying and the humane element in a scientific education.

INVENTIVE THINKING IN THE PRIMARY SCHOOL

EDWARD DE BONO

The Cognitive Research Trust, Cambridge

The three intellectual ages of man seem to run as follows:

0– 5 years old	the age of WHY?
5– 10 years old	the age of WHY NOT?
10–75+ years old	the age of BECAUSE.

If you ask a child up to the age of 10 or 11 to draw a space machine he will invent his own space machine—usually a sphere with legs and compartments for such essential commodities as sandwiches and a box for moon samples. After the age of 11 he will be more likely to draw a conventional rocket with NASA written on its side. Inventiveness has given way to imitation. It is easy for an adult to put a higher value on inventiveness than on imitation because an adult knows how things should be and so he can indulge in the luxury of inventiveness. But a child who is exploring the world puts a higher value on finding out how things really are and demonstrating this knowledge.

So young children are inventive for two very good reasons. They do not have a chance to be anything else—since their knowledge is limited. They enjoy thinking and playing with ideas. A child is not original in his thinking because he knows the conventional approach and has chosen to go beyond it but because he simply does not know the usual answer and has to put one together for himself from his experience.

A child's thinking is idea centred rather than information centred. It is only later that thinking becomes object centred and then very much later (and only with sophisticated thinkers) that it becomes process centred again. A child is not really interested in a ladder as an object but in what a ladder does. It is a pity that we so often teach from objects to their function instead of from functions to objects—which is the way children learn naturally.

How can the inventiveness of a child at the primary-school age be encouraged? It is not enough to say that children are naturally inventive and given a generally stimulating environment they will practise their inventiveness. It is necessary to set tasks and ask them to do things. I once asked some children whether they had ever had any good ideas. One girl of 12 said that she had had good ideas *three years* before when she had been asked to design some things by me (via a series I ran in an education journal). It seems that in the intervening three years no one had ever asked her to come up with an idea. So it is necessary to set specific tasks and projects: design an elephant-weighing machine; show how you would stop a cat and a dog from fighting; draw a dog-exercising machine, etc.

When I first started getting children to tackle such tasks many adults (both teachers and parents) told me that they were surprised at the reaction of the children. They had expected the children to ask 'what for?' or to have difficulty in getting started. Instead the children would accept the request as perfectly natural and would at once dash off a dog-exercising machine or whatever else was requested.

What is of the utmost importance in this age group is the medium used for thinking. Language is quite inadequate because it is vague and indefinite and cumbersome when it comes to describing complex relationships or mechanical devices. Furthermore language is very dependent on the home and cultural background. It would be absurd and counter-productive to expect the children to construct actual working models of their inventions. Not only would they lack the tools and technical ability but they would be restricted to small-scale inventions and it would take much too long a time. There is no doubt in my mind that drawing is the perfect medium for practising thinking in young children. By drawing I do not mean the artistic portrayal of 'scenes' like cottages and cows but inventive or problem-solving drawings like dog-exercising machines and elephant-weighing devices.

A drawing can be done quite quickly: 10 minutes to half an hour. There are no technical limitations and most children can draw well enough for the purpose. With a drawing a child can express an idea—*and the assembly of ideas to achieve a purpose*—in a much more definite manner than with words. He must link things up. He cannot say vaguely that 'somehow it happens'. He expresses what he can express and then uses black boxes and 'special switches' to cater for the processes that are beyond his understanding. Once finished a drawing can be looked at by teacher and child together. They can pay attention to specific areas: fill in gaps, change and improve the design, etc. A drawing can be compared to other drawings and points of similarity and difference can be discussed. It was for all these reasons that I chose drawing as the basic medium for the 'Thinking Course for Juniors' (5-12)* which provides thinking practice in invention and problem solving.

The problems set by the teacher must involve things which interest the children and also come within their range of experience (second-hand experience will do). So ice-cream, chocolate, cars, elephants, dogs, policeman, trains, etc., can be used. There should be a definite purpose that has to be achieved and plenty of action. Descriptive scenes are useless. The task set can, however, be quite complex or even abstract, e.g. how to stop a cat and dog from fighting; how would you deal with bad men?

In summary it may be said that children in the primary school age group are naturally inventive. They love thinking. They love playing with ideas—to achieve a purpose. But they must be set specific tasks in order to develop their inventive skills. Drawing is the ideal medium.

* Published by Direct Education Services.

INVENTIVE THINKING IN THE UPPER SCHOOL

ERIC LAITHWAITE

Professor of Heavy Electrical Engineering, Imperial College of Science and Technology, London

If I can see further than others, it is because
I stand on the shoulders of Giants.

Isaac Newton

How many teenagers, already fascinated by science and the multiple inventions of modern technology, have bemoaned the fact that they were *born too late*? 'If only I had lived in Faraday's time there would have been so much for me to invent, and so much of it was easy!'

Among the most difficult obstacles to modern fifth and sixth formers are Newton's Giants. They were few: Leonardo and Galileo were undoubtedly two, and much use he made of their work. How very many more have sixth formers today: how can they ever hope to read and appreciate the work of Niels Bohr, of Geiger and Marsden, Max Planck, Einstein and Ernst Mach, work for examinations and still find time to experiment and to build on their work?

The one thing that might convince sixth formers that it is still 'all there for the picking' is as yet denied to them, for it is unlikely that they have made a scientific discovery for themselves. Had they done so, they would have known that only yesterday the problem seemed like the North Face of the Eiger. Yet today it has crumbled into a heap of dust so small that you wonder how you ever tripped over it. And tomorrow?—tomorrow it will seem as if the solution had been known to you from your earliest school days.

Somehow they must be persuaded of this. Since there is no sure method by which this might be done, teachers could therefore adopt a truthful approach and themselves say 'I don't know' in reply to a teenager's probing question, and resist the temptation to be evasive on the assumption that 'someone *must* have done something on this simple question'. It would help too if teachers themselves actively engaged in their own bit of research, however trivial. The great electrical engineer Professor F. C. Williams of Manchester University had a quotation typed and framed: I guess it still faces all who enter his office and sit in the visitor's chair. It is a quotation of a former Principal of Owen's College and reads: 'He who learns from one who is himself engaged in learning drinks from a running stream. But he who learns from one who has learned all that he teaches drinks the green mantle of a stagnant pond.'

We cannot all become as inventive as Professor Williams but all those who have the courage to want to see their pupils become better scientists than themselves can at least follow him by *doing* rather than thinking, and by not being afraid of making mistakes.

Inventiveness is not the sole property of the scientist or mathematician. There is a seed of it in all of us, and it could just as easily blossom in the brewer (like Joule) or the bookbinder (like Faraday). There can be few of us, if indeed any, who never said an original thing or thought of something that was his or her own. Nor is originality the prerogative of the male sex, either, no more than is science itself.

How in the interstices of time spent on preparing for 'O' or 'A' levels can one find room for a little extra curricular activity? I suggest that there is no better starting-point, nor one occupying so little time, than a visit to a Patent Library. Take out any volume of British patents for a year between 1840 and 1890 and just begin turning the pages. Out of this exercise can come many things such as:

1. A marvelling at the ingenuity of man.
2. That essential humility referred to earlier.
3. The realisation that many of the so-obviously 'good' ideas never bore fruit, for commerce and science do not necessarily go hand in hand.
4. Most of all, the discovery of a new process or a new material makes commercially feasible that which was not so in 1869.

Take out the old ideas and dust them and be amazed at what shining pieces of gold you might find. What about a day trip to a Patent Library for a small group of fifth or sixth formers followed by a short write-up of what they found that 'turned them on'? (If nothing, then nothing need be written.) The results could be read out to the rest of the group—you have the makings of your own 'Think Tank'. Short of this, let teacher alone make the visit and devote just one period to relating the day's experiences. It may have no relevance to 'A' level passes, but it might be a key in a lock that might open a door to a new world.

Above all, never underestimate the value of individual thought. This is no place for false modesty or inferiority complexes. This is something to be excited about, to share and to be enjoyed. Today we go into Space, but still we cannot escape the solar system. Tomorrow we shall cross the Universe—and if we don't it will have been fun trying!

SIMULATION: IN THE CLASSROOM

JOHN TAYLOR

Head of Town and Country Planning, Trent Polytechnic

Simulation techniques are centuries old; it is only their importance in the classroom that has lately become more widely recognised. Over recent years at least three major streams of simulation have built up a significant reputation and each of these deserves specific mention.

Role-play is a simulation technique which relies on the spontaneous performance of participants when placed in a set situation. It is undoubtedly the simplest form of the three types of simulation considered here, as it usually has little formal structure. All that is required is that the participant accepts a new identity, in effect he steps inside someone else's shoes and acts *or* reacts appropriately. Thus in role-play the essential core of the activity is understanding another person's point of view in a constraint-free environment. Case studies can be used for role-play by participants being presented with a real problem and real data and then being asked 'What would you do next?'

Gaming is a simulation activity which is slightly more complex than role-play because it relies on more formalised procedures and a greater structuring of relationships. Academic games customarily consist of groups of players placed in prescribed settings where their actions are in part governed by methods of procedure and various rule systems.

Most games are primarily concerned with a desire to understand a decision-making process involving both quantitative and qualitative elements.

Machine or computer simulation is an activity which is rooted in mathematics. Probability theory, game theory and other associated mathematical techniques are utilised to build chance and random elements into operational representations which depend upon the electronic calculator or computer to monitor dynamic situations. The purpose of machine simulation is usually to find 'answers' rather than to understand processes. Here the emphasis is on quantitative elements which can be handled by the machine and often human participation is limited to an initial set of decisions and a response to a final result.

The distinction between these three types of simulation is not clear-cut and definitions in this rapidly developing field are still a matter of considerable dispute. There obviously can be a large degree of overlap between role-play, gaming and machine simulation; a simulation game, for example, can be constructed to contain a role-play; it can also require the use of a calculator or computer to expedite its procedures. In the present context, finer distinctions are relatively unimportant as these notes concentrate on simulation games to highlight some of the virtues of the technique and some of its major operational features.

In considering some of the advantages of the technique two themes are particularly prominent. One set of 'pay offs' relate to the development of student motivation and the other collection of claims cover the immediacy of perceived relevance.

Motivational advantages can be listed as follows:

(a) Heightened interest and excitement in learning.
(b) The divorce from the 'conventional wisdom'.
(c) The removal of traditional student–teacher polarization.
(d) The familiarity of a universal behavioural mode.

Gains related to relevance and learning can be listed as follows:

(i) Simultaneous learning within a single game can take place at diverse levels.
(ii) A range of decision-making experience over differing time spans can be readily offered.

(iii) Students can directly appreciate other roles and vantage-points by stepping into someone else's shoes.
(iv) A synoptic view can be presented as a vehicle for free inter-disciplinary communication.

Set against these claims it must be achnowledged that though simulation activity is growing fast it is an educational approach which is still in its infancy. In short, users should expect to encounter 'teething problems' related to the limited amount of experience achieved with particular models. It must be acknowledged that simulations are time-demanding activities, especially if packaged material is rejected in favour of a custom-designed approach.

Finally in setting down these brief reservations it must be clear that using simulation in the classroom is not a short cut to easy success. A demanding skill is required and without this expertise perhaps more harm is done than through not using the technique at all!

In acquiring a competent mastery of the technique it is helpful to be reminded of the fact that most instructional simulations set out to produce their desired effects through:

(a) presenting a simplified abstraction of the bare essentials of a situation free from trivia and irrelevance;
(b) concentrating on making explicit essential relationships and the fundamental interplay between key roles;
(c) unfolding time at a very much quicker rate than normal so that the implications of action in a dynamic situation can be clearly and repeatedly felt;
(d) allowing students to 'sit in the hot seat' and feel the direct impact of the consequences of decision-making;
(e) offering opportunities for collaborative learning on self-directed lines (i.e. learning as much from one's mistakes, and from the mistakes of others as from one's success).

Keeping these factors in mind can be helpful in realising some of the technique's full potential. Clearly much work remains to be done if even a small part of the technique's early promise is to be properly realised in the immediate future. It is to be hoped that this text, along with others, will encourage a wider audience to acquire an operational understanding of the technique prior to personal involvement in discriminating experimentation and everyday usage.

References

Taylor, J. L. and Walford, R. *Simulation in the Classroom*, Penguin Books: London, 1972.

SIMULATION: SOME MODELS

MICHAEL BASSEY

Reader in Education, Trent Polytechnic

1. You are the leader of a neolithic community of twenty adults and a number of children farming in a clearing in an oak forest. You have decided to move a large limestone boulder across the clearing and to raise it onto six smaller boulders. How are you going to do it? (Later generations will come and stare at your structure and call it a megalith. They will describe the larger stone as weighing 19 tons and the smaller ones as 1 ton each; they will say that you moved the large boulder 100 yards. Others will measure it in tonnes and metres, others in frons and uugs.)

Every school in Britain has a megalithic monument within reasonable travelling distance and every class could measure the stones, estimate their weight, calculate how many men with what levers and rollers could have built it, and make drawings and models which the visitors would admire at open day. Worthwhile as the experience would be, it misses some of the vital elements of technological sensibility; there are more important questions than "How are you going to build it?"

Who decided to build it? (Autocratic or democratic decision?) What will you do if some of the others decline to help? (Persuade or enforce?) If you need more helpers how will you get them? (Hired labour or slaves?) Who will till the soil and care for the children while you build? What damage will your operations do to farmland or to people? How will you pay your helpers, or feed your slaves? Why are you making it this shape and size? What will you do when it is finished? . . . Why are you building it?

Too often these political, economic, moral, aesthetic and spiritual questions about technological achievements are forgotten.

2. You are Dr. Martin Brown, 25, recently obtained Ph.D., unemployed for 3 months since leaving university ('nobody seems to need my science skills'), married to a teacher who is at present the breadwinner, wanting to start a family, prepared to live anywhere. You have been offered an appointment as . . . . *, well paid, good amenities, reasonable prospects and a stimulating challenge to your ability.

But the work has attracted adverse comment. Your wife puts arguments to you suggesting that this is not a proper way of using your scientific talent. A Dr Smith, with whom you were friendly as an undergraduate and who would be a colleague if you accept the appointment, puts arguments opposing your wife's view. Do you accept or reject the appointment?

3. You are one of the councillors of the planet Clarion which has just been colonised by 100,000 humanoids; you will be in office for 50 years and your brief is to develop the planet while maintaining a reasonable quality of life for its inhabitants. Humanoids are susceptible to holobinkitis, a fatal disease for which there is at present no cure. Mining of walterite is the most significant feature of the economy. You are asked to draw up a budget every 5 years and the consequences of your decisions are relayed to you.[1]

4. You are . . .

. . . a railway pioneer developing a route between Derby and Manchester in the 1840s[2];

. . . a company man engaged in the planning and operation of a fleet of coastal tankers[3] or of North Sea gas and oil exploration[4];

. . . a director of a subsidiary of Hemlock Industries and every time you take a 'socially responsible' decision your salary is cut[5];

* Quality control chemist of a napalm manufacturing plant in the U.S.; or member of a research team developing non-lethal harassing and riot-control agents; or technical sales manager of a company manufacturing polychlorinated biphenyls; or scientific adviser to a company manufacturing aerosol sprays.

. . . involved in a public enquiry into the siting of a new power station[6] or of a new air terminal [7];

. . . a member of a River Board planning the cleaning of a polluted system of Scottish rivers[8];

. . . a member of an Aid Committee deciding on the allocation of limited resources to development projects in Botswana, Upper Volta and the Windward Islands[9];

. . . an adviser to the Commission for the Environment of the European Community during the period from 1975 to 2000 and dealing with a range of energy, resource and social problems[10].

5. You are one of a small group of survivors of a world-wide catastrophe and living in a 'survival cell' which is linked to the outside world, monitored, and controlled by a computer—Humanus—which communicates to you through a voice print-out on tape. Humanus requires certain decisions of the members of the cell if it is to continue to survive[11].

6. You are one of a group of people who are going into hibernation for a period of 100 years. (By Cryogenically Halted Metabolism technique.) You have been invited to give what the tabloid press is calling 'The Methuselah Lecture' on television and there will be millions of viewers throughout the world listening to you. You have just read *Report from Iron Mountain* ('War is itself the principal basis of organisation on which all modern societies are constructed') and *Blueprint for Survival* ('The principal defect of the industrial way of life with its ethos of expansion is that it is not sustainable'). What message have you for the world?[12]

Simulations like the last three are dangerous. A well-conducted simulation provides an experience closer to reality than any other educational strategy and herein lies the danger of simulations which examine technological change in a future context and by social, moral and political criteria. Talk to people who are technologically well informed and who have studied both the growth reports of the Hudson Institutes and the no-growth reports of the Club of Rome—or similar writings. Some of these people are pessimistic about man's future and may be both cynical about efforts to influence our future and hedonistic about the present. Teachers who use effective methods for bringing people face to face with the problems of our times should not be surprised if they swell the ranks of the pessimistic, the cynical and the hedonistic. But hopefully they will also increase the numbers of those who are prepared to use their creative talents morally and aesthetically to meet the challenge summed up in Gandhi's epigram that there is sufficient in the world for each man's need, there is not sufficient for each man's greed.

References

1. Showalter, V. M. *Planet Management Game*, Houghton Mifflin, Boston, U.S.
2. Cole, J. P. and Smith, G. A. *Railway Building*, Department of Geography, University of Nottingham.
3. Atthill, C. R. A. *Decisions 2—Coastal Tankers*, School Government Publishing Co. Ltd., Bletchingley Road, Merstham, Redhill, Surrey.
4. Walford, R. *North Sea Exploration*, Longmans, London.
5. Bassey, M. *Corporate Hemlock*, Department of Education, Trent Polytechnic, Nottingham.
6. Falk, N. *Laceville Power Station*, CRAC, Bateman Street, Cambridge.
7. Lewis, D. and Carson, S. McB. *The Conservation Game*, Hertfordshire Education Committee.
8. Falk, N. *Star River Project*, CRAC, Bateman Street, Cambridge.
9. Oxfam Education Department, Oxford.
10. Bassey, M. 'European Environment 1975–2000', Conservation Trust, 12 London Road, Chertsey, Surrey.
11. Twelker, P. A. and Layden, K. *Humanus*, Simile II, La Jolla, U.S.
12. Bassey, M. 'Teaching resources on the social responsibilities of management'. *Management Education and Development*, **6**, 1, 32–40 (1975).

SIMULATION: A CASE STUDY

J. K. JACQUES

Industrial Science Department, University of Stirling

Bertrand Russell wrote that the task of the philosopher was, primarily, *to ask the right questions*. If we are to be successful in unravelling complicated problems then, whether we are philosophers or mere scientists or sociologists, we shall need as many tricks and devices as we can lay hands upon to guide our questions.

These tricks and devices are chosen from the range of modelling techniques listed elsewhere in this book. In selecting and using models, simply remember:

that: (a) they will only answer the question you have asked and

(b) wisely used, they will illuminate particular aspects of your problem rather as a plan view or cross-sectional drawing will emphasise only those features relevant to the blueprint in question; they will *generate better questions.*

To illustrate these vital points and particularly to provide a logical framework for the teacher to operate in, here is a compressed version of a case study which has been used very successfully with fifth- and sixth-year pupils.

The idea for this case came from newspaper accounts of a fire in a block of flats in a city suburb. The numerical information used to generate the appendix material came from published government reports and Building Research Laboratory statistical pamphlets; and I would strongly emphasise the wealth of such material which is readily available in lending and reference libraries throughout Britain.

The Case Description

Phoenixton is a busy industrial town, close to the Metroburgh conurbation. Currently its fire engines are provided by a branch of Metroburgh's central fire service, about 5 *miles to the north-west of the town.*

Last year there were three serious fires causing the loss of four lives and over £2,000,000 *worth of damage. The worst incident was the destruction of a warehouse of synthetic fabrics, where two employees were trapped and burned to death; in another tragic case, two young children died of burns because of an unattended paraffin stove in their parents' flat.*

These disasters led to much adverse criticism in the press about the long time taken by the fire engines to reach the scenes of trouble. Others said that criticism of the fire service was unfair, that braver firemen could not be found anywhere, and that the real trouble lay in the maze of narrow, traffic-congested streets—a planning problem unlikely to be solved within the decade.

The Town Council has now appointed its own Fire Chief—Mr McFarran, formerly of the Metroburgh Fire Service. A government loan of £2m *can be made*

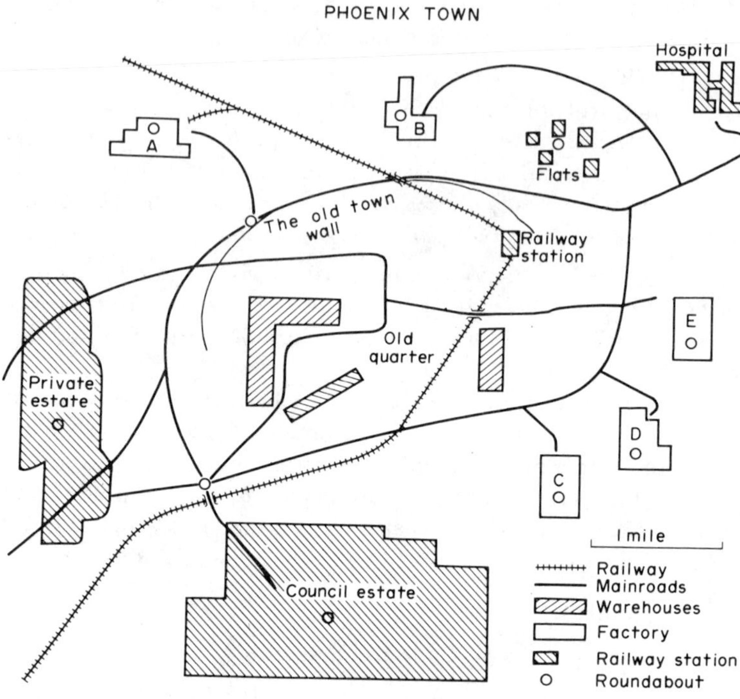

Figure 1.

available for the construction of a fire station and the provision of fire engines and the Town Council has voted £0.3m per annum for running the new service.

You are asked to act as advisers to Mr McFarran on all aspects of the problem. Groups of four to six pupils are given the case description and then tackle the 'problem' by using the following 'System Modelling Check List', or something similar.

Boundaries (How wide is the area for study? The system here is Phoenixton and its people.)
Constraints (e.g. existing road networks and housing estate layouts).
Inputs and Outputs (Commuter traffic? Goods on lorries, etc.? Effect on street congestion).

Major Properties in Phoenixton and their Values

Property	Estimated value of property and contents	No. of people living in or working at	Statistical risk of an outbreak of fire in any one year	Estimated chance of more than 50% damage if there is a fire
	£ million			
Factory A (chemicals)	10	800	1 in 20	0.15
,, B (glass)	2	178	1 in 100	0.30
,, C (weaving mill)	1	350	1 in 30	0.40
,, D (golf balls)	3	57	1 in 200	0.20
,, E (needles)	0.5	46	1 in 100	0.10
Flats (north west)	1.2	800	1 in 10	0.01
Shopping area (central)	0.5	140	1 in 75	0.03
Hospital (NE)	2	160	1 in 20	0.005
Council houses	2.2	700	1 in 20	0.01
Private estates	3.8	800	1 in 30	0.01
Warehouse complex	6	90	1 in 10	0.06
Railway station	0.2	10	1 in 100	0.001

Time Variables (Effect of rush-hour traffic patterns, factory working hours, etc.).

Sub-systems (e.g. the separate housing estates, factories, traffic-island arrangements).

Relationships between Sub-systems (Do they help or hinder each other, or are they quite independent?)

Optimising Functions for Sub-systems (The 'importance' of the chemical factory may demand a separate, internal fire engine!)

Information Networks (Telephone, fire alarm, TV surveillance?)

What are the Problem Centres? (Will different people see the problem differently? Yes!)

What Alternative Solutions are Available?

Optimising Function for the Whole System (Public and private sectors may not be 'congruent'—resolve conflicts here!)

Performance measures and controls (Build models to test whether proposals are sound, i.e. hypothesis testing!)

The Problem of Dealing with Conflicting Optimisation Function

This throws into very sharp focus the fact that students may not really have grasped the deeper nature of the question 'what *is* the problem?' Is it about fire-engine accessibility at all? Is it about prevention of fire, or about the time for 'messages' to reach fire stations? This involves design of buildings as well as design of street and telephone networks!

Within this problem area we must disentangle questions of risk, human values, social and private costs of fire and injury.

How good is the information available? (Always test your pupils' critical faculty by providing some 'irrelevant stuffing').

How practicable is the proposed set of solutions? Always test the hypothesis!

Tackling the Problem

As they begin to tackle these questions, the student groups will demand information—the map and the statistical summary provided here are merely examples of what can be provided at the start, or withheld until requested.

Students may wish to interview fire officers; or they may wish to test ideas on 'laboratory bench models'. Very simple card models might be constructed to study spread of fire through rooms or corridor complexes (construct on an asbestos or steel sheet and set fire in an open space on a windless day!!), or to study the response times of simple fire detectors.

The distribution of fire station services in the town can be argued through with the aid of a hard-board mounted map of Phoenixton.

Weights (proportional to 'fire importance' of the site) can be suspended on light cards through holes in the board; a sliding ring on the board attached to these cards will come to rest after suitable coaxing at the (sub)optimal position for the fire station. Or several separate rings and card systems can be used (see Ackoff and Sasieni for development of this idea).

References

Gwilym M. Jenkins and P. V. Youle. *Systems Engineering: A Unifying Approach in Industry and Society*. (Watts & Co., 1971, 'The New Thinker's Library'). The best general introduction.

de Neufville and Stafford. *Systems Analysis for Engineers and Managers*; McGraw Hill, 1971. Z20.7. A good development reading, *after* Jenkins and Youle.

Ackoff and Sasieni. *Fundamentals of Operational Research*, Wiley. Especially the early chapters on models.

R. N. Blair and C. W. Whitston. *Elements of Industrial Systems Engineering*, Prentice Hall, 1971. Emphasises human activiti es as well as non-human.

R. F. Miles. *Systems Concepts*. (Lectures on Contemporary Approaches to Systems), J. Wiley, 1973. Z20.7. An amusing discoursive analysis on some modern technologies—reflects American thinking.

H. Chestnut. *Systems Engineering Tools*, John Wiley, 1966. A sound, clear introduction to mathematical and computer tools.

A. G. J. MacFarlane. *Engineering Systems Analysis*, Harrap, 1964. Z20.1. A more advanced but very clear text.

Open University Texts

Designing the Future, T 262 1 O.U.

Technology and Society, T 262 2–3 O.U.

Using Linear Models, T 341 3–4 O.U.

Relationships within Models, T 341 5 and 6 O.U.

AWAKENING THE TECHNOLOGICAL SENSIBILITY OF GIRLS

ELIZABETH HALSALL

Senior Lecturer in Comparative and Secondary Education, Institute of Education, University of Hull

It seems astonishing that girls have acquired the reputation for not being technologically minded. They used fire and water for cooking, they pounded maize or cassava, they did all but the heaviest work on the family's food plot and were responsible for the spinning of cloth. All domestic technologies require skills of hand and eye, as well as brain, just as joinery or engineering do, though perhaps not of the same kind.

Whence comes the idea that women are not technologically oriented? It looks as if the Industrial Revolution in its earliest manifestations may have been partly responsible, with its emphasis on factory work and heavy machinery. For though a girl of moderate gentility could and did spin at home, the factory, with its heavy machinery, was unsuited to their muscular skills.

Many women worked in textile mills where their capacity for fine manipulation was in demand. In lacemaking too they did well, but manoeuvring a capstan they were less good at. Still a vast number of women now drive cars and, contrary to previous masculine impressions, they have a lower accident rate than men. They jack up and change tyres and some even do courses in motor-car maintenance.

The matter repays examination from another angle, that of the knowledge available about the respective skills of males and females from the findings of physiological and psychological research. Girls hear better than boys, boys see better than girls. Girls have rather better linguistic skills than boys, boys have markedly better visuo-spatial abilities than girls. Girls have a better sense of smell and touch than boys and incidentally they are more sensitive to pain. They are also better at fine manipulative skills. Girls are therefore better in industrial tasks requiring swift and dexterous assembly of small parts. They are also better at typing. Boys are better at numerical, mathematical and mechanical skills and at the interpretation of mechanical information pictorially represented. Being more muscular, they are better at heavy tasks. Girls are better at practical tasks, being aided therein by their better rote memory which enables them to remember for short periods a number of unrelated items not personally relevant to themselves. This skill also makes them better at clerical and executive tasks.

Boys are more object-oriented, girls more person-oriented and each sex is more capable of greater complexity of behaviour according to its own particular orientation. Boys show greater variability, for good or ill, than girls.

Many of the above differences are not linked to particular environments or cultures but appear at least in embryo from soon after birth, say, by the age of 10 weeks or so. The particular sex-linked skill is often reinforced by the mother. For example, girls' early babbling elicits a response from mothers who talk back and therefore talk more to infant daughters than to infant sons. Some of these sex-linked differences are not confined to humans but are found also in, for example, chimpanzees. These various skills, though sex-linked, are not exclusive to one sex. There are many individuals in the other sex with high scores on skills untypical for the sex. The writer's sister, for example, when on the beach can knock tin cans off the tops of pebble cairns with greater accuracy than any man she has ever met. Her husband, who has learnt to live with the phenomenon, gets great amusement out of the squirmings of her unwary male victims.

A quick perusal of the above prompts the reflection that, whereas in earlier technological developments, particularly in the 'domestic' technologies, girls were no more handicapped than boys, the first Industrial Revolution and the Scientific Revolution preceding it required great numerical, mathematical, mechanical and visuo-spatially angled skill and, in the case of the former, great muscular skill of a gross kind except as regards spinning and weaving.

The second Industrial Revolution seems to recognise these feminine attributes. Electronics requires greater manual dexterity and greater practical capacity, for example, in checking readings on unrelated dials.

Yet in spite of the work of Dame Caroline Haslett* it seems a waste of

* Dame Caroline Haslett's work along two fronts was of a pioneering and seminal nature. Originally an electrical engineer she worked to secure use of electricity in the home, founding the Electrical Association for Women in 1924 with this aim in view. By the 1950s it had 1400 members, mainly housewives and teachers. She also worked to

time to train girls in the appropriate skills when they are likely to be given the chance to operate only at a lower level of skill than boys.

We may best get at the right answer to the question of training girls for more technically oriented skills if we turn the question on its head. It is well known that a technological world demands literacy from its workers and citizens. It is true also that boys have poorer linguistic skills than girls. Should we then not bother to make them literate, but leave literacy to girls? Watch every male hackle rise: perish the thought. We have remedial classes for illiterate boys. Let us have technical classes of any kind for any sort of girl. For all children we must certainly build on their strengths—and girls have manual dexterity—but we ought not to neglect their weaknesses. Since women use mechanical instruments in the home and drive cars outside it, since they have 20 years of working life after child-rearing is over, they need some technical training in technologies other than the domestic.

How then can we sensitise girls to the need for such training? In the first place by ridding them (and boys, since adolescent girls will not do what their male peers think to be unfeminine) of the idea that technical work is unfeminine (and that domestic technologies are not masculine). For such attitudes are a far bigger handicap to adolescent girls still searching for their identities and their roles than minor weaknesses in technical aptitude. How can we then convince both boys and girls of the need for girls to develop technical skills and understanding? By placing the above facts before them in some form suited to their capacity to absorb them. For the facts of physiological and psychological skill and the facts of surviving stereotypes are both at odds with present needs. Let us point out in particular that hands that are good at lacemaking and embroidery are equally capable of being good at gem making or at electronics, with its miniaturised components or, for that matter, at eye or heart surgery. Women can always leave the heavy technology—or the butchering—to the men!

---

secure the admission of women to careers in engineering and to appropriate examinations, in line with her conviction that admission to such careers would help to raise the whole social status of women. She was the first secretary of the Women's Engineering Society and editor of its journal *The Woman Engineer*, and the only woman member of the Electricity Authority set up in 1947. Every woman—or man—who ever pushed a Hoover owes her an immense debt of gratitude.

At the practical level schools should be encouraged to undertake work of the kind done at Walkden. At this school the high motivation of average girls for domestic subjects was tapped both to sensitise them to the technological approach and to improve domestic equipment. Similarly, their great capacity for caring was tapped to produce aids for the young, the old and the disabled. At the same time care was taken to ensure that such projects were not conceived as purely the province of girls, so that the attitudes of boys towards domestic technologies improved. The exciting work done at Walkden merits imitation elsewhere and should prove a stimulus to radical thought and approaches to the whole problem of sensitising girls to technology and changing the attitudes of both sexes about girls' roles in the modern world.

Brief Bibliography

Association of Teachers in Technical Institutions. *The Education, Training and Employment of Women and Girls*, new ed. A.T.T.I., 1973.

Hunt, Corinne. *Males and Females*, Penguin, 1975.

Jenkins, E. W. The scientific education of girls since 1902. *Durham Research Review*, no. 32 (Spring 1974), pp. 873–86.

Lewis, K. L. The science of homecraft. *School Science Review*, **54** (Mar. 1973), pp. 429–40. (On the scientific background to home-making skills.)

Llewellyn, B. Technology for girls. *Studies in Design Education and Craft*. **5**, no. 1 (Winter 1972), pp. 51–4.

Mead, Margaret. *Male and Female*, Penguin, 1970.

Miller, Ruth. Careers girls don't choose. *Where*, no. 105 (June 1975), pp. 158–60.

Woodhall, Maureen. The economic returns to investment in women's education. *Higher Education*, 2, no. 3 (Aug. 1973), pp. 275–99.

Ways and Means

PROBLEM-SOLVING

PHILIP ROBERTS

Curriculum Adviser and Head of Design Department, Manor High School, Oadby, Leciester

A familiar term: yet, if teachers examined implicit assumptions so often made regarding the nature of it, there might result not only new opportunities for pupils but also fresh insights into how learning proceeds and what supports and environments are most appropriate. This kind of development activity, undertaken by teachers in their own schools, is the essential element in curriculum change: without it, 'curriculum packages' are of no avail.

For some people, a problem is defined by a precisely expressed and measurable objective, often in terms of an artefact: the making, say, of a table or hovercraft. The definition of problems in terms of precise, material objectives puts the answer before the real question. Can it mean that the conditions are so well understood that the only valid responses are in terms of a table or a hovercraft? Does it mean that a great many alternative responses have been considered and rejected? If so, does there exist some methodology of learning that has made the criteria of choice and the process of pupil decision sufficiently explicit to the pupil? Is a problem to be seen in terms of making an impersonal product to an abstract formula?

I suggest that in technological education in schools we should enable pupils to locate themselves in their real situations, in and by which they must identify real problems, identify individual and social, moral, ethical and economic needs and values, as *they* distinguish between criteria and constraints, in their route-finding to acceptable change (or not). The implications of this are radical: there is a world of difference between problems found (educationally too much solved) in books, and existential problems. It's a move from How? to a very hard Why?

If problems are stated in terms of impersonal artefacts, their solution is often seen in a sequential, step-by-step process. Whether such a process assists in the significant development of even cognitive skills may be questioned. Real problems, however, are usually autobiographical and social. The way in which they are identified expresses those shifting relationships that enable change and personal growth to take place. The problem represents an evaluation of the discrepancy between 'what is' and 'some condition other than that'. The methodology is the recognition and the reconciliation of the discrepancy. The acceptable closure of the gap is the solution. Herein lies a fundamental point. It is not solution in the sense that we have now finally reached the definitive conclusion but rather that there is a better match between insight and skills. In this process of expanding insights, pupils' awareness of discrepancies between 'what is' and 'some condition other than that' is in constant evolution. Consequently solution becomes re-solution, and growth is seen in the re-solution, over and over again, of pupils' needs and aspirations.

Educationally, impersonal problem-solving leads to the idea that if you *can* do something technologically, then you *ought* to. Skills of know-how then become out of step with wisdom. On the other hand, a real problem-solving approach would, as part and parcel of its methodological validity, insist that resulting technology be congruent with, and responsive to, people's values. In this view, problem-solving is towards greater responsibility and interaction between people and place, and the recognition and the acceptance of responsibility for diversity, confrontation and uncertainty.

We must necessarily arrive at our own procedures, appropriate to our own schools, and the useful material can be found within some of the newer design methodologies; in operational research, cybernetics, value analysis, ergonomics; in research on change processes and innovation and creativity; in synectics which, for instance, makes a useful distinction between the problem as given and the problem as might be understood; from artists and poets; from philosophers of science, say, Popper, Kuhn and Polanyi; from such as Carl Rogers, Abraham Maslow . . . all available.

But I should be sorry if this sounds theoretical and beyond our pupils. In practice it is neither. Nor is such activity only possible with clever children. It is necessary to remember that in a book we use words to describe and that description can never be an equivalent of action. All that our reading can do is to offer insights.

However, there is another trigger that might be useful in developing a procedural approach and that is to pose the question: to what problem is this process or artefact or argument or plan a possible solution? Apply it throughout the studios and workshops and school, and it can raise fundamental questions and transformations. Experience it essentially with colleagues before beginning with a group of pupils because it is impossible to develop a sound pedagogic approach towards problem-solving unless those responsible for it have first-hand experience themselves. Or practice your developing methodology in, say, staff-room issues because those are real problems. Consider How? and Why? and what happened. If nothing changed then the consideration of that is valid in its context. Notice that all the time we are thinking of process in action whether we label it learning, changing or technology. If staff-room issues are too close for comfort, work on those that fill the local paper. Consider that problems are *now*, they are *here*, and they are, more or less, urgent. Why and how they transform can tell us of change in process.

As *we* change in our changing context, we can reflect on our own learning or changing, and then work towards the provision of the most appropriate conditions—the relationships between people and place—in and by which pupils change. In so doing, we shall find out a great deal about ourselves and our pupils. And we shall, incidentally, be making a constructive contribution to the shaping and reshaping of society whether from an educational, political, technological, ethical or epistemological point of view.

NEW RESOURCES

ARNOLD MILLER

Education Producer, B.B.C. Radio Humberside

The most sophisticated operation is to re-create an image from the printed word. Especially if that printed word synthesises the documents of a preceding generation. Yet this is precisely what a high proportion of teaching about man's past imposes upon students of all ages. Small wonder that some of our great past technological strides spark off something less than proud amazement and respect in so many classes.

To minimise these barriers to understanding film and television can do much, but from the teacher's point of view these represent remote techniques whose products he must accept or reject largely as they stand. The good teacher will be on the look-out for worth-while material from these sources, but at the back of his mind he will always want to generate his own resources, tailored to the needs of children he knows, and possibly drawing on the local environment to which he and they have ready access. Perhaps best of all he will want to do this alongside those children with whom he is working. For all but a lucky few this will preclude the use of film and television, and so this article is about successful experience with other means.

It is relatively easy, and now quite common, to by-pass the synthetic textbook summary. Archive teaching units of various types have presented a selection of relevant (facsimile) primary sources with the intention of evoking some kind of more direct feel for the past. So far so good. Contemporary pictures, diaries, handbooks and prospectuses will convey something of the dynamic of past struggles for progress. But in many cases the most stimulating exercise has already been done, namely the process of gathering, assessment and selection. The object of the Museum Studies experiment launched at Danum Grammar School in 1969 (as CSE mode 3S) was to enable children to make their own researches and selections from among local original documents housed at Cusworth Hall Museum.

Conditions for such a project could hardly have been better. The archives and exhibits available were a rich source of information on South Yorkshire. The curator, who owned much of the collection, was not only willing to make it all available to young and inexperienced researchers, but was also generous in providing ample working space, and his own invaluable guidance. It is heartening, and perhaps a lesson for more nervous museums, that he did not find it necessary to curtail these privileges.

The method of working was simple enough. In preparatory sessions several topics from social and industrial history were selected and the foundation knowledge absorbed. At the museum the relevant documents and exhibits were examined by small groups of boys allocated to each subject. A dozen or more centrally significant documents were chosen for reproduction by offset lithography, photocopying or electric stencil cutting and the study unit was completed by a work-guide derived from group discussion of the available material and its place in the story of local progress. Thus canals, coalmining and railways were not rather static and remote economic developments, but became the explanation of local work and communication patterns. Similarly the themes of law and order, nonconformity, and trades unionism were seen as part of the way of life of students' own past relatives. Powers of analysis and evaluation had been demanded by the selection and editing process. An awakened lively interest in local social and industrial developments were the result. This was reinforced by visits to and slides of sites illustrating the themes of each unit as it evolved.

Once complete, such a carefully constructed study unit is of permanent use, at least as a foundation for project work, and possibly also as a springboard for further enquiry using other sources and other methods. One such simple but largely under exploited aid is the tape recorder.

These machines have been with us in some schools for 20 years. Their function has been oddly circumscribed throughout this period, confined as most of them are to the recording of material originated elsewhere. Its role as a simple but effective means of capturing verbal impression and memory or of staging vivid illustrative dialogue without the need for excessive rehearsal remain mostly unexplored. Sadly in those instances where some attempts have been made, the results are often marred by easily avoided pitfalls—the lugubrious interviewer, the bathroom acoustic, the stilted stylised conversation. It may be that the mystique and reverence surrounding the skills developed by broadcasters have left us with a widespread reluctance to experiment.

The development of local radio stations from 1968, each complete with an integral education department, provided the opportunity for teachers to acquire and exploit techniques that previously either baffled or eluded them. Within 3 years of the setting up of BBC Radio Humberside, for example, eight teams of teachers were regularly involved in the production of radio programmes and associated support material. There were between eight and sixteen in any team—hence up to eighty teachers involved in taming a previously unpredictable medium for their own needs. They learned to write scripts skilfully, with all the economy and precision in the use of words that that involves.

They learnt to interview effectively, probing forcefully or sympathetically and with disciplined sense of purpose. They even learned to cope with the confusing display of studio controls to produce programmes that were widely regarded as of high quality, and worthy of the standards set by their professional broadcasting colleagues. And why did they succeed so consistently so well? Because they were using skills in the preparation of programmes that they used daily in planning lessons. What did they make? Dramatised incidents from local archives, to enliven our appreciation of the Grimsby Dock Strike for example; a radiovision study of the River Hull, involving a range of officials that no school could hope to invite, and producing a bargain filmstrip; themes from literature emanating from war, illustrated by memories of local people who experienced similar trials to those immortalised in poem or prose; a cacophony of sound to stimulate primary-school children in their local project work. All local teachers, gathering local material, to answer a locally felt professional need.

Teachers proved capable and willing practitioners. But children of school age surpassed them. They too made programmes. They interviewed with a candour and facility that amazed most of us. Guided by a teacher who had made some effort to master these techniques, they could add a new dimension to their study units. There are still men who sailed our canals. There are men who can describe pit conditions better than any official account, men who drove steam, men who struggled to establish unions. From these and the other sources could spring a dramatised tape slide sequence made by the class itself. With what benefits? The excitement of the actual pursuit of knowledge unrefined by an intermediating pen; a sense of direct contact with our achievements by meeting the evidence that has survived; an ability to search out information and subject it to thoughtful comment—all privileges normally enjoyed by the one who produces the standard book summary, assuming that he too has not missed them. This must be no special preserve! Nor need it be.

TECHNOLOGY IN THE MIDDLE SCHOOL

DENNIS F. MARSHALL

Adviser in Design, Craft and Technology,
Doncaster Metropolitan County Borough

What makes the shop door open when a person approaches it? How do the traffic lights work? Who switches on the street lamps at dusk and switches them off again? How does the toaster know when the bread is toasted? These and many more questions are asked by young, curious children observing their everyday technological environment. How well can the schools answer them, or help them to work out the answers for themselves?

The early experiences of children, building with bricks, playing with model cars and railways, the imaginative play on the beach with sand and water must be further developed in the schools to capitalise on the child's natural curiosity, ingenuity, imagination and motivation. Children should be involved actively and excitedly in the technology around them so that they come to realise a number of comparatively simple parts which are easily understood, and in this way acquire the confidence to accept technological achievement rather than be frightened by it as, regretfully, many teachers are. The experience provided must be enjoyable if an inquiring attitude is to be developed and continued throughout life.

When a child wishes to compare the weight of two objects, why produce a balance? Why not pose this as a problem for the child? Metre rules and elastic bands are available in most classrooms and there may even be some springs. What of the child who makes use of a piece of wire and a violin bow!

Much middle-school project work develops from the local environment, but how often is the technological aspect developed? The building site, the farm or the docks provide many opportunities but rarely do we see working models which could extend the project in a most stimulating way.

On a windy day a teacher might ask the children to estimate the speed of the wind and the direction from which it is blowing. The range and effectiveness of the children's solutions would probably amaze many teachers. Often, because of their limited experience, children attempt and succeed in developing ideas which the adult would invariably dismiss as impractical. The children not only enjoy the constructional work but because moving parts are involved, have to measure accurately, acquire new skills and set themselves high standards. The evaluation brings not only fun but also a critical assessment of each of the various solutions.

This experience might lead to the development of the same theme. How can the wind be used? Kites, sailing boats and wind-driven vehicles could be made and tested. What a range of problems to overcome! A visit to a local windmill might be arranged, a thorough study of the mechanisms made and a working model constructed. This might appear to be a very ambitious project for middle-school children but some schools have shown that complete success can be achieved. Gliders, helicopters and parachutes could also be explored.

Maybe the use of windpower to propel a vehicle gives the opportunity to discuss energy sources and introduce the idea of storing energy in an elastic band. The cotton reel tank is the traditional solution to the elastic-band-driven vehicle but children are capable of much better ideas than this and the distance moved can be measured in terms of metres, not in centimetres, much to the excitement of the children.

Devices which operate automatically fascinate children. How does the record player pick-up arm choose the correct position to lower on to records of differing sizes? This operation can be observed, the possibilities discussed, and afterwards the mechanism examined in detail to establish the facts. But many automatic systems can be made. Using simple components such as reed switches, micro switches and bi-metal strips children can make fire and burglar alarms and vehicles that stop automatically, for example.

For a real test of ingenuity, children could be asked to devise a means of communicating a message across the playing field when it is not possible to make oneself heard or to communicate to a deaf and blind person.

For this type of work to be successful the school must provide children with suitable facilities, the relevant opportunities and adequate time.

A wide range of materials is required. Constructional materials such as Lego, Meccano and Fischer Technik are extremely useful particularly since electric motors are available to drive models. Why involve children in very difficult constructional work if simple, easy to use constructional methods are available. However, a great deal can be done with paper, card, empty containers, string, elastic bands, wire, paper clips, etc., and lack of expensive facilities need not prevent a start being made.

The classroom environment can be enriched by the introduction of an ever-changing museum of items of technological interest. A collection of locks, clocks and simple mechanisms which children can dismantle and explore. Simple working models which encourage investigational work might well be included. For example, the provision of a simple water wheel could involve the pupils in an investigation to compare the efficiency of the overshot and the undershot methods which would require a high level of ingenuity and accurate measurement. Interest aroused might well lead to a detailed study of Smeaton who was himself involved in similar problems.

Children should be encouraged to use tape recorders, cameras and film projectors to gain the confidence to use technological equipment.

The child who is late for school because his bicycle tube punctured, or chain came off, provides an opportunity to examine closely the bicycle—an excellent example of simple but effective technology. Teachers should remember also that many children help parents service the family car and are quite familiar with its working principles.

Comparatively few teachers in the middle schools have adequate background to undertake this type of work with confidence. Interest must be developed by showing what can be done, and through in-service courses teachers must be encouraged to become involved and helped to recognise the opportunities which exist. Even the experienced and knowledgeable teacher, however, will find that he will not be able to answer all the questions and must be prepared to listen to and learn from his pupils and to work alongside them on occasions, a situation some teachers find difficult to accept.

CREATIVE TECHNOLOGY AND CONTROL SYSTEMS

GERALD FOX

Danum Grammar School, Doncaster

Not so many years ago few teachers in secondary schools had first-hand experience of the project approach to learning. Most had come across the term, some had a little knowledge of its implications while others rejected the idea as unstructured and inefficient, and doubted its educational value. Those who had read some of the literature or had seen the approach at work in some schools are not necessarily to be blamed for being sceptical. Let us be realistic, an unstructured learning environment inevitably leads to inefficiency and is likely to fail to maintain the interest of both student and teacher alike. Consider the student who has an initial interest in a topic which he wishes to pursue further. He sets out full of enthusiasm but quickly becomes aware of inadequacies in his knowledge of the subject. Progress is also slowed by seemingly endless construction. All too frequently the end result is a device which remains incomplete, works very badly or not at all. All too frequently the teacher has to intervene to try to salvage the foundering project.

How does one avoid disasters of this kind? In the 'open' form of project just described the major problem appears to be a combination of shortcomings in basic knowledge and the time required for construction. If construction forms a major part of the project, it must be started early. Yet this is the very time when the student has the least knowledge of his topic. This paper outlines one possible way out of the difficulty by considering one field of technology but the solution has more general application.

Implicit in the concept of the project is that the student uses his creative ability in trying to solve his problem. A great deal of literature has been written on the various aspects of creativity though much of it is devoted to argument about definition. Frequent references are made to the 'supposed' creatives; to Einstein, to creative scientists, to creative architects, and so on. But we need not limit ourselves to such high-level creativity. Everyone has a creative potential and the teacher can provide the environment through which each student can reach towards his ceiling. Ausubel (1963) has referred to this 'lower' level of creative ability as a 'general creative ability'. By this definition, therefore, it is valid to nurture the creative ability in all our children.

But even the advocates of the 'open' form of project could claim this for their approach. However, Wallach and Kogan (1965) have reminded us that 'if we assess a person's capacity to generate cognitive elements, one factor influencing his performance as a ceiling or upper bound is the extensiveness of his repertoire'. Thus it is fundamental that we ensure the student has the relevant repertoire before he sets out on his project—not the answers but the background knowledge which will enable him to use his ingenuity in finding them. One method which has been used is to ensure that, before a student begins his project, he reads a number of specially prepared cards relating to the area of enquiry. The limitations of this approach, however, are fairly self-evident. Few teachers have the time, incentive or ability to prepare reference cards and information sources for even a limited number of areas in technology. How does one foresee what area a particular student is likely to favour?

A more practical method is to select an area of technology which will lend itself to a wide variety of problem situations, appropriate at school level, and to prepare a number of 'basic course units' which will provide the requisite background knowledge. One such area is control techniques and the writer and a colleague have developed the Control Technology Course to this end*. A basic course introduces students to a number of control techniques from which they can later select to solve their own problems during project work. Students become familiar with the relevant principles through a series of investigations rather than through instruction, and in order to allow for one group of students

* See Schools Council (1974a,b,c) in the References.

working at a different pace from others, a programmed series of assignments are used. An assignment completed, students refer to a Follow-up sheet which indicates some possible solutions—certainly not 'the answers'—and further information on the topic.

There are other advantages of using the programmed assignment method. Not only are students able to work at their own pace, they are able to undertake the investigations to differing levels. Staff are freed from the traditional role of the teacher. They have the opportunity to visit the different groups—groups of three are typical—to discuss their work and to provide, repair or even design equipment. Should the teacher wish to introduce new material, he simply writes a new set of assignments and Follow-up sheets.

As distinct from the investigation, the project implies a certain amount of constructional work. In the open project—one without explicit structure—much educational value is lost through excessive time devoted to construction, particularly if the initial design is suspect. Clearly, unless steps are taken to prevent it, the introduction of a basic course will add to the burden. It is essential, therefore, that special equipment should be designed such that most of the time is spent in thinking about a solution and a minimum in its fabrication. For instance, the electrical equipment used in the Control Technology course consists of 'boxed' units, such as resistors, capacitors, diodes, relays and transistors with their connections brought out to 4-mm sockets. Structures are assembled from Meccano and later from materials such as Handy Angle. Basic equipment should not only be reliable, it should have the possibility of being quickly assembled and be attractive in appearance.

Given that students have adequate background knowledge, through a basic course or otherwise, one must not overlook the possibility that they may be quite unfamiliar with the project approach. Not only are they required to solve a problem, they must also be able to perceive the problem in the first instance. Indeed, it is rarely a question of the one problem and the single solution. It is important that the student should recognise that there is a problem, that there are several problems to be solved in a major problem and that there are several possible solutions to any single problem. The experience which helps to bring about this recognition is gained if students attempt a number of minor projects at intervals during the basic course. Not least they should also have become aware of the necessity of keeping an accurate record of the procedures used.

Some years ago I had occasion to examine the written accounts of both investigations and projects produced by students in the same school. In the former accounts the students had used existing equipment to investigate phenomena, whereas, in the latter, students had set out to solve a major problem which involved constructing equipment. The difference in the quality of the two kinds of account was quite staggering. The accounts of the investigations were clearly and methodically written, but without exception the project accounts failed to make clear what had been done and, in general, would have been of little use to anyone wishing to obtain information for a similar project. It has since been found that by encouraging students to record their work during the basic course, coherent accounts of a project have become the rule rather than the exception.

Having completed a basic course students will be equipped to produce a successful project. Should they work individually or in groups? Unfortunately examination requirements have tended to dictate the answer on the grounds that group projects fail to identify the individual contribution. Yet if we ignore such constraints for a moment, it is apparent that the number involved is likely to be a function of the project itself, or of the individual student. Some may wish to work individually, some may prefer to work with others—'brainstorming' can be helpful—or the magnitude of the task may necessitate a whole group of, say, fifteen or so students working towards the solution. Should we not adjust the examination to suit the course rather than vice versa?

An associated problem concerns the acceptability of a project. If the course is, say, primarily concerned with control, must the project involve the control of a system? Is it permissible to allow a project if it makes use of the techniques learned during a basic course but otherwise has little association with the title of the course? These are the kinds of question teachers have asked in relation to Control Technology. Some insist that the project must be concerned with control, others that control is not essential if the basic course material is used extensively in the project, while others are prepared to accept virtually any area of technology.

What teachers should be asking themselves is what is their primary aim in introducing their course. If the aim is to give expertise in a particular area of technology, then one would expect the project to be closely associated with that area. If, however, the primary aim is to encourage a 'creative' technology, then provided that the student possesses an adequate background knowledge in his area of interest, we are probably not justified in denying the student's right to proceed. For instance, Control Technology was devised because it provided a wide range of problem situations, not because it was considered to be the most important area of technology. It is suggested, therefore, that it is unjustified to be unduly pedantic about the actual content of a project following a basic course unless the content of the course is of primary importance.

It is a fairly straightforward matter to ascertain whether or not students have learned the basic principles and facts from a course in technology. It is far more complex to determine whether students have become more 'creative' as a result of a course, unless we make certain assumptions. If we accept that a divergent thinking ability (Guilford, 1956) and a lateral thinking ability (De Bono, 1970) are essential to the creative technologist, then the problem becomes a little simpler. We can then rephrase the question to ask ourselves whether a course in creative technology has produced an increase in these two abilities. The results of a recent piece of research are quite encouraging in this respect. Fox (1975) found that students who had taken a course in Control Technology had a significantly greater divergent thinking ability and lateral thinking ability than a random sample of students from the same population who had not taken the course. In addition, these superior abilities were evident both within and without the context of technology. It is suggested here, therefore, that in order to achieve successful and imaginative project work in schools, there must be an adequate foundation, perferably through a basic course, involving the knowledge of basic facts and principles, the encouragement of 'creativity' and giving experience in a methodical approach to project work.

References

Ausubel, D. P. (1963). *The Psychology of Meaningful Verbal Learning*, New York: Grune and Stratton.

De Bono, E. (1970). *Lateral Thinking: a Textbook of Creativity*, London: Ward Lock.

Fox, G. J. (1975). An evaluation of Control Technology: a course in technology for secondary school students of average and above-average ability in the 13–16 years age group. Unpublished M.Ed. Dissertation.
Guilford, J. P. (1956). The structure of intellect. *Psychol. Bull.* **53**, 267–93.
Schools Council (1974a). *Control Technology Teachers' Handbook*, London: English Universities Press.
Schools Council (1974b). *Control Technology Pupils' Assignments*, London: English Universities Press.
Schools Council (1974c). *Control Technology Pupils' Follow-up Sheets*, London: English Universities Press.
Wallach, M. A. and Kogan, N. (1965). *Modes of Thinking in Young Children*, New York: Holt, Rinehart and Winston.

OPENING WINDOWS ON TECHNOLOGY

KEVIN WALTON

Adviser on Engineering matters, Malvern College

"*Dear Mr. Professional engineer,*
"*I am writing as Headmaster of Elmbury Grammar School to ask if you would take part in a careers evening for sixth formers. We would allocate you a room and you should be prepared to answer questions from pupils about engineering as a career.*"

I wonder how many schoolmasters have written such letters and how many engineers have taken part? But more than this, how many have wondered just how much has been achieved when the occasion was over? Speaking as a schoolmaster I am sure that such evenings do good and speaking as an engineer, they are certainly better than nothing—but only just!

If I were asked what is meant by Engineering, I would prefer to indicate what it isn't. It is certainly not 'applied science' and the life of an engineer has little in common with the life of a scientist. Yet to a school pupil and most schoolmasters they are all bracketed together. Engineering as a profession is far wider than this. An engineer is involved with identifying and proposing solutions to real problems. He requires a peculiar mixture of curiosity and imagination, creativeness and the logical power of mathematics, the ability to express himself, and an appreciation of the possibilities of science. Judgement, humility and human understanding are all part of his life.

These sort of thoughts combined with the Profession's increasing awareness that enough quality pupils of high potential are not entering the Profession have prompted the experiment 'Opening windows on Technology'. This is but one way of getting over to pupils, parents and school staffs alike what Engineering is all about.

The problem is to find out how to attract more high-quality pupils into the Profession. 'Opening windows on Technology' is only one of the solutions. To say that an engineer is 'born not made' has much truth in it, but how easy is it to identify the born engineer? My observations in schools, and colleges, in industry and Engineering, suggest that helping to locate the 'born engineers' is the field in which the practising engineer can be of most help. Somewhere, sometime, before it is too late, a pupil must get a glimpse of what is involved in being an engineer and measure himself up against the sort of person that he should be and this at a time when he can still see visions and dream dreams.

Once the vision has been seen and the dreams dreamed, the normal process of academic education can proceed. Some schools are traditional, some are untraditional, some specialize in broader general studies, some specialize in expression and art. All of these aspects of education fit into the profession of being a good engineer.

But how to start such dreams? This in a personal way is what I have attempted to do: to find by experiment the right age group and find by experience the things the dreams are made of and how to get these things over; to find the sort of people who can conjure up such dreams and to find a way of triggering off such dreams. This cannot be put over second hand and such people must come from the Engineering profession itself. Engineers turned schoolmasters tend, regrettably, to be fugitives from the Profession: the opening of windows must be performed, if it is to be successful, by missionaries. They must be practising members of the Profession and seen to be so by the pupils of the schools they visit.

A pilot scheme has now been started with this sort of background. With funds from the Institutions of Mechanical and Civil Engineers, some fifty firms up and down the country have been asked to nominate 'up and coming' engineers, mostly under the age of 30. Each of the nominees is asked to search his engineering career for something which can be found to illustrate what Engineering means to him; some problem capable of clarification and definition which he or she has had to tackle and resolve and take to completion; something which shows some of the processes of thought required of an engineer and which demonstrates the difference between the scientist, the applied scientist, the technician and the engineer. These nominees after training and trials in the classroom would be available to visit schools, just to 'open the windows' and help the pupils to look out. They are not intended to be careers lectures and there is no attempt to by-pass the job of the careers master.

Experiment has shown that visiting engineers are often far more effective if the classes are small. It is better to take 5 classes of 30 pupils than to attempt a large single lecture to 150. Such lectures can achieve very good rapport and have a better chance if they are treated rather as current affair discussions than as part of any science syllabus. It is better to give such lectures too early rather than too late. The sixth form is too late; specialisation has already started. The pre-O-level years are probably the best. Since lectures should fit into the school time table they should not last more than 40 minutes and with practice this is found to be possible. Topics will have no particular attachment to any syllabus and will in fact be seen to link up with all subjects. All modern geography, for instance, includes water and water supply in some detail, but it is the engineers who are involved in building the dams, drilling the tunnels and working out the complicated distribution system which will fill our country's need. A town centre in Northumberland was the architect's dream, but it was the engineer who had to do the historical research in the British Museum to find records of coal working before he could start the foundation and organise the labour and do the work.

The pilot scheme that is now running, 'Opening Windows on Technology' can be seen to supplement all the other aspects of the teaching of Technology, but it has the advantage that it can provide a service for the large percentage of schools where Technology has no platform and Engineering is still mixed up in the minds of pupils with overalls, nuts and bolts, dirty hands and making things work.

If these nominated engineers do this job well, they can do nothing but good; but if they do not, they can do untold harm. The pilot scheme ensures that people who go into the classroom with this brief are properly trained, are attractive personalities and capable of putting over in a fair and unbiased way what Engineering means to them. These are not lectures on the scale or style of the Leonardo da Vinci lectures of the Institution of Mechanical Engineers or the TV lectures at the Royal Institution. The aim of those lectures is to paint the broad canvas of the wonders of Engineering. The aim of the lectures in 'Opening Windows on Technology' is to build up a rapport between the pupils and the visitor so that the youngsters say: 'I like this chap, he obviously enjoys what he is doing, he is not all that much older than I am, I could in fact be that sort of person myself in a few years' time if I set my mind to it.'

We as engineers can no longer sit back and complain that pupils are not entering the profession. We must be seen and be heard to be doing something about it. 'Opening Windows on Technology' is doing just that.

DRAMA IN A TECHNOLOGICAL SOCIETY

ROGER DAY

County Drama Adviser, Wiltshire Educational Authority

Hilary Johnson, who is 16 and a winning essayist finds that Technology has already 'wrecked not only nature but the experience of being alive'.

If Hilary had been involved in an exciting and well-run drama programme at the school she attended I feel sure she would have questioned further what technology is, or is not, doing for our society. For our young to survive and live life to the full in an ever-increasing technological world, it is essential that they positively believe that the experience of being alive is valuable, and that they contribute and control some areas of their existence. Why should such questioning result from an involvement in drama? After all, theatre in the seventies is making vast use of the mechanical and electronic resources now available. Electronic lighting equipment, hydraulic stage lifting devices, closed circuit television in numerous areas of the auditorium, uses of new materials in stage design, metals, polystyrene, polythene, etc., new processes in the dyeing and cleaning of costumes, all are being utilised to explore and develop theatrical presentation.

To understand how Hilary could have come to question and reason the values of technology, I wish to ignore the term 'theatre' and concentrate on the term 'drama'. 'Theatre' to me means a polished art form, incorporating sound, light, colour and movement, and presented before an audience. 'Drama', as I see it, means a group experience, exploring in a creative way a particular idea, or set of ideas, not necessarily presented before an audience. Its purpose is purely for the benefit of the participants; they not only enjoy the experience, but they gain knowledge through the learning situation. By working as a member of a group involved in a drama experience one becomes aware that everyone, no matter what their capabilities, has something to offer. In accepting everyone's contribution, however small, it will add to and change the shape of the creative process. Working together like this allows the group to put themselves in 'others' shoes' and to see how it feels; also to examine many potentially dangerous areas in the safe situation of a drama space.

Drama then is the process of living through, and exploring, any given situation. This ability to share and exchange ideas, to see how one's particular emotions, and thoughts, can be interpreted in many ways, is most important in our technological age. Man has been made increasingly aware that there is always someone, or something, to do things for him. There is someone else, not only to create the problems, but to solve them for him, too. He can, therefore, feel that he need no longer relate to other individuals because so many mechanical devices do the relating for him. This inability to relate causes tremendous problems, and is leading more and more people into isolated lives.

In industry many people have found that better results are being achieved by workers when opportunities are given them to work as a group, to present a finished product. The individual alone, doing one particular task, not seeing how his or her particular job affects the whole, will hopefully become a thing of the past. The ability to communicate verbally with others is most important, if people are to work like this. However, it is an ability that is still not encouraged to the full in our educational system. Many teachers still place the written word on a higher pedestal than the spoken word. I feel that there should, and must, be equal opportunity for both. In all walks of life, verbal reasoning and communication is as necessary and as important as being able to move and breathe. To think, place ideas logically, and communicate them to others, must be useful in an industrialised society. If, however, these functions are taken over by machinery, man will become slave to an automaton.

With drama in Education touching hopefully more and more young people of all abilities, each will think about their tasks in a new way; and not only will they become aware of themselves, they will become aware of others, which is of prime importance in any kind of society. Drama then can lead people to question, and share, common and not-so-common experiences. Perhaps it can help close the circle of living that technology leaves open, and so solve Hilary's dilemma.

Case Histories

IN A GIRLS' SCHOOL

MOLLIE BARRETT

Science Mistress, Walkden High School, Manchester

A Scientific Peep into a Kitchen of the Year 2000

Alongside the traditional Science course, Project work can be fitted in as time and opportunity permit. Thus at a girls' secondary school at Walkden, a life-size model of a kitchen of the future was made by the girls for a Science Fair. Before building it, they studied current technological trends and developments, and tried to forecast their effect on the pattern of living by the turn of the century. In each case we checked back with the industry concerned, to ensure that the items of equipment represented in the kitchen were technically viable.

The demountable model, 6 feet high and 11 feet wide, was made chiefly of wood and hardboard, and each one of the 250 girls in the school played some part in its construction. To use a saw, a brace and bit or a spanner was a new experience for most of them, but what they lacked in skill was more than compensated for by enthusiasm and ingenuity. They enjoyed lifting and manoeuvring the large and cumbersome hardboard panels, shouting instructions to get them to fit, aligning holes to receive the securing rods, and bolting and screwing the equipment in place.

We must be vigilant that in the mixed School situation such activities do not become the sole prerogative of the boys.

As their teacher I am interested in work study and have always encouraged its practice in school, especially that aspect known as time and motion study. Eventually it was given a formal place in the time table, and for 3 successive years the girls manufactured Christmas crackers as an exercise in the use of work study techniques.

The Household Brush

In 1963 the Royal Society's Committee for the Encouragement of Scientific Research in Schools accepted an application from the school to carry out research on 'Work Study in the Home'. Our first project was an investigation into the design of the household scrubbing brush. After examining and testing more than 100 scrubbing brushes the girls recommended an improved design, and their report was published by the Council of Industrial Design. Our second project was an investigation to find the best method of storing dried fruit in the home, and the findings were published in the journal *Packaging*.

In each case, publication resulted in contacts being established with organisations outside the school, opening up opportunities for further purposeful work. We investigated biscuit tins for a well-known firm of biscuit manufacturers, and subsequently a tin designed to the school's recommendations appeared in the shops. An account of the school's work was read by a productivity officer in the Punjab, and this led to an exchange of correspondence between our pupils and a school in Amritsar.

Feminine Fashions

A project 'Science Around the Stiletto Heel' dealt with the widespread problem of the pitting of floor surfaces by the then current fashion in ladies' shoes. The girls studied the effects of the stiletto heel on different types of floor surface, and devised equipment to investigate various aspects of the problem. One class visited the local zoo to find the weight of an elephant and the area of its footprint. They were then able to prove the truth of a statement that the pressure of an 8-stone woman on a stiletto heel is 50 times that of an elephant standing on one foot.

Aids for the Young, and the Disabled

The girls devised anti-baby-snatch equipment, consisting of a number of devices which could be fixed on or near a pram to give warning on the approach of an intruder.

Apparatus to help the deaf included alarm clocks adapted for switching on a flashing light to provide a wakening device for the deaf. The girls then suggested the simple yet effective idea of suspending from the bedhead a hair dryer connected to an electric time-switch, so that, at a predetermined time, a current of air was directed at the sleeper's face. Amongst other equipment constructed was a wind-strength indicator, and a door-bell system giving a visible instead of an audible warning.

As the school's work became known, people began sending us their problems. Our attention was called to the difficulty experienced by blind people in assessing the level of the liquid they are pouring into a container. After months of work we developed a liquid-level indicator which gives out a 'bleep' when the probe comes in contact with a liquid.

The Appliance Officer of the Spastics Society enlisted our help on behalf of a severely disabled teacher in Sussex who was unable to write on a blackboard, and we developed for her a means of visual communication.

An Automatic Distress Warning System

A device developed by the school to provide an automatic distress warning system for people living alone is now in commercial production. I had read a series of distressing reports in newspapers of elderly people who had collapsed through accident or illness lying helpless, sometimes for several days, unable to summon assistance. All the warning systems available at that time required manual operation, and were therefore useless in situations of greatest need such as sudden or unexpected collapse.

It seemed surprising that no one had thought to devise an automatic system, using, say, photo cells and electronic circuitry. We lacked both the knowledge and the facilities to use these sophisticated techniques, and set to work with such limited skills and equipment as we possessed.

After almost a year's work and many frustrations and disappointments, we had produced a viable piece of equipment. It consists of a time-switch mounted on the side of the lavatory tank, so that each time the chain is pulled the time-switch is returned to the 10-hour mark. If a continuous period of 10 hours passes without the lavatory being flushed, the time-switch brings into operation a warning light or buzzer located at the front of the house.

This device received favourable notice on radio and television and in the national press. During the next few months more than seventy local authorities and other organisations concerned with the welfare of the aged wrote to ask for further information. Two facts became increasingly clear, that there was a widespread social problem, and that our device was the first attempt to provide an automatic solution.

As a result of initiative taken by Manchester Corporation, the device is now available commercially, and a number of local authorities including Manchester have installed pilot schemes.

I would like to feel that the work carried out at Walkden illustrates not only that girls, but that girls of average ability, can make a purposeful and worth-while contribution to the technological society if opportunities suited to their interests and their capabilities are offered to them.

IN A PRIMARY SCHOOL

SIDNEY SMITH

Headmaster, Rodbourne Cheney Junior School, Swindon

The tedious task of motoring up and down a stretch of the M4 running in a new car at a sedate 40 mph eventually led me into the north section of the Leigh Delamere Service Area for refreshment. Its 42 acres, spread over both north and south sides of the M4, with its well-planned layout, impressed me. As I watched the comings and goings of the huge juggernauts and private cars over the top of a plastic cup, the idea of a project based on the Area began to take shape—What have we here to stimulate the lively minds of our 10–11-year-olds? A variety of skills to be acquired through Maths, English, Science, History and, judging by the heavy foreign lorries, we could well take care of their geography for the remainder of the term.

Here was an opportunity for young children growing up with a motorway running through their county, to learn something of the needs of the lorry driver and the motorist, of how far these needs were being met. This would oblige them to find out in a practical manner, acquiring skills on the way, in a field of learning related to their environment.

The teachers would be able to demonstrate how the traditional subjects of the school curriculum could be enriched and made relevant to aspects of life outside the confines of the classroom.

A brief survey of the Area resulted in the following points for consideration:

1. Petrol pumps, vehicle service area and recovery.
2. Car parking area.
3. Lorry and coach parking areas.
4. The cafeteria.
5. Employment of staff.

This seemed an ideal situation in which to involve our two fourth-year unstreamed classes, and the idea was discussed with the two teachers concerned who, I was relieved to learn, expressed a keen interest in the proposal and readily agreed to start planning.

First doubts appeared after a preliminary visit to the area by the two teachers; that of safety and supervision of the children, as huge juggernauts belted into the parking area, engines revving and belching black smoke! There was also a service bridge over the M4 giving access to both north and south complexes of the Area, which could well cause problems, to say nothing of the fairly fast movement of private cars, coaches and vans to and from the parking areas and petrol station.

But we agreed that the five areas of investigation would suit group study by the children and set about convincing the Head Office of Granada and subsequently the General Manager of the site. This was effected. So the staff paid a second visit to Leigh Delamere where they found the employees most helpful and genuinely interested in our scheme; no section of the complex was barred to us and we were even given access to the kitchens!

It was thought that each of the two classes should spend one day a week for six consecutive weeks at the site, one on Tuesdays and the other on Thursdays, and a coach was booked for the whole period. In spite of setting off at 9 a.m. and returning to school at 4 p.m., we found the working day limited to roughly 10.30 a.m.–3.15 p.m. with a break for a packed lunch. To keep the necessary timed recording operations continuous, staggered lunch breaks were made for those children concerned.

As the Service Area occupied both North and South complexes of some 42 acres and supervision and liaison would be extremely important, it was arranged that both class teachers together with our Junior Helper would accompany each class whilst I took the class remaining at school.

A scale map of the Service Area was going to be a 'must', and the County Architect was contacted who was able to let us have two 1:500 copies. We were able to supply most reference books from our own school library such as history of roads, bridges, road transport, fuel, etc. We thought it would be useful if we could fit in some comparisons between this motorway and those of other countries. The West German and Italian Embassies were most helpful in sending us glossy expensive books on their own networks of motorways.

The backbone of investigations, however, came from workcards and worksheets prepared by the teachers, and the thorough preparations and discussion in the classrooms prior to the visits.

The problem of coping with children who would be working in both North and South areas separated by the service bridge and about half a mile distant from each other was solved by dividing the children into groups of approximately three, each responsible for contact with one teacher centred at each complex, with one adult having a roving commission. Each group knew where to go for equipment—or help—and teachers knew where each child would be (or should be).

The morning of the first visit of each class was used for observation, children familiarised themselves with the layout of the area, got to know their work-points, met the staffs of the cafeteria and petrol station and came to understand the few rules we thought necessary.

We met the local M4 Police Patrol who was often in the area and put him in the picture. He was able to play a very useful part in our scheme back at school at the end of our project.

It took the last 3 weeks of September to complete preparations, run off work sheets, questionaires for interviews with the public, sheets for recording petrol sales, etc., centralising our library of reference books, collecting a variety of measuring instruments and stop-watches, and making sure the two portable tape recorders were ready. Children were encouraged to have their cameras ready for use and the school's super 8 movie camera was well stocked with film to make a record of the project.

As the operations would be taking place during October and the first half of November parents were advised as to suitable clothing for their children.

In the field, the co-operation between children was wonderful. Unusual or unexpected incidences which would spark off new avenues of exploration were immediately accepted by them and encouraged by the teachers. Back at school they would follow their new line of discovery and thus build up an entirely new field of work which may not have originally been thought of. Individual aptitudes and skills quickly became evident, and it was rewarding to see these being acknowledged and accepted within and between the groups of children.

It would take more space than I have at my disposal to write in detail about the work which evolved from the project so I will condense this under the various headings.

Petrol Pumps–Vehicle Service Area

Children worked for half-hourly sessions recording gallonage and grades of petrol sales, including amounts per customer, using the pump meters. They even, on their own initiative, did a time and motion study on the pump attendant, working out the distance he walked over a timed period!

One interesting fact relating to probability arose when graphs were being drawn during class session at school. There were five different brands of petrol, each brand established on one of the five islands (in line at regular intervals) to which cars could draw up. Children found that the highest sales came from the 'X' pumps. The question they asked was: 'Did motorists prefer 'X' petrol, or did the position of the 'X' island have any bearing on the result?'

Of the various ways of testing this, two are worth mentioning. Five chairs were placed at equal distance apart and children from various parts of the school were invited separately to sit on one and so on. In the second method one lad took a 4-foot length of board, screwed on five tins at equal intervals and filled each with milk straws. Children were then asked to walk along and take a straw from any tin. This again was done so that the following children didn't see what the previous child did. The 'X' chair and tin recorded top sales!

The children found out the position of the underground storage tanks, their capacity and shape. Exercises on capacity to scale, etc., followed. Incidences of breakdown and main causes of such were also investigated. Measurements of the Service Station were made for further use of scale models later.

Lorry and Coach Park

This proved an exciting, interesting and at times an amusing field of study. Children interviewed drivers on arrival: where they had come from, their load (not always revealed) and their destination, how long they were away from home, why did they choose to drive a lorry, why did they stop here, did they always do so, and their opinions of the area. A note of the makes of lorries and their power was also recorded.

Two incidents are worth recording in a little detail. The child who sought the help of the lady member of staff. 'Miss —, we can't understand him and he's dressed so funny!' It was discovered that driver and lorry had come from Marseilles with a cargo of shell fish for Bristol. A return journey of 7 days, said the driver. Armed with the route of the journey some interesting work developed back at school.

The second incident concerned another foreign lorry, this time from Denmark. The Compass Rose trademark of the haulage firm was the prime mover here. A Danish driver who spoke English. A cargo of tinned meat. Any samples? asked the children! Do you use English drivers? Yes. Do they go to Denmark? No! Why? They can't speak Danish!

We got the address of the firm and the children wrote to Denmark. This began a very interesting section of work from the information sent to school from Denmark.

Car Parking Area

The tape recorder played a main part here. We used our extroverts who tried to keep to their prepared questionnaire, but very quickly this was disregarded. There were things they wanted to know which perhaps we had overlooked. The children wondered if motorists having had a long tiring journey would welcome an overnight stay. This proved very popular, and generally the answer was 'Yes, we hadn't thought of that—it would be nice—sort of a Motel.'

The children began to realise the importance of the Motorway to the commercial traveller. He could see home more often, a quick meal at 2.30 p.m. and he is back in London for tea.

Children began to take notice of those drivers who wore seat belts, something we had overlooked, and this was logged.

The tape recorder was invaluable, not only recording interviews to be sifted and evaluated back at school, but to superimpose on film, together with the general sounds of activities going on all the time.

We wanted to map the numbers of vehicles from various counties. The children blew up a large map of the British Isles from the overhead projector and from their recordings were able to plot with tape the volume of traffic from each county. This proved very rewarding. It gave the children some idea of where the counties were and their size, and led them to investigate why large volumes come from certain places whilst from two counties, no vehicles were recorded.

The Cafeteria and Shop

Here again the tape recorder was our main aid. Customers were interviewed by children and readily told them what they thought of the food, the service and the cost.

During the periods of investigation the management decided, owing to shortage and cost of plastic utensils, to change over to more orthodox ceramic cups and saucers, etc. This proved a change for the better with customers.

Some very useful and practical maths work came from the kitchen, particularly from tea and coffee sales where certain inquisitive children having found out the amounts in weight of these ingredients for a given number of cups, were able to estimate the profit margin. They were able to observe and time the 'automatic sandwich maker' and various other time- and labour-saving devices. Needless to say, the children engaged in this section usually took back home half their packed lunch!

Employment of Staff

One child wrote: 'Leigh Delamere is like an oasis set in a double ribbon of road surrounded by Wiltshire's beautiful countryside, miles from anywhere. How did the people who worked there get there? Were they dropped by helicopter?'

The children learnt that the Area gave work to practically the whole of a nearby village. They were given transport to and from the centre. But didn't the farmers lose some farmworkers? Oh yes, but the farmers also lost some farmland.

Recording Traffic Flow

Groups of children were engaged in recording, from the connecting bridges, types of traffic using various lanes both east and west bound. We needed this information to graph traffic volume throughout each daily session, and children were relieved every half-hour. From research the children did on past annual increases in volume of traffic and the estimated future increases, they were able to plot what the position might well be in 10 years' time. A resulting scale model of a section of the M4 showed two and three storied cars and lorries!

In conclusion I think the reader can well imagine the wide range of follow up work that took place in the classrooms. Children spent 5 afternoons a week for 6 months. Maths, geography, history, science, model-making, creative writing, painting and drawings all found their rightful place within the framework of the scheme.

We began with the safety aspect and ended with it when a police sergeant of the motorway patrol arrived with the Range Rover in the playground one afternoon. He spoke at length to the children about his work and showed them the drill of laying out signs and the equipment used in emergencies. This comprised the last 20 feet of our filmed record —the Range Rover's lights flashing, siren wailing and eager children, inquisitive to the last, seated behind the wheel.

IN AN UPPER SCHOOL

JOHN JEFFERY

Pocklington School, Pocklington, York

Appropriate Technology in the school situation has, for me, two quite distinct aspects. The one is concerned with involving pupils directly in tackling technical problems of the 'Intermediate Technology' type. At its best this type of project can provide great scope for imaginative work and possibly lead those involved to a greater understanding of the problems of the developing countries.

My experience of these projects suggests that if the latter is the principal aim, then it is unlikely to be achieved. Not only is it difficult to find projects suitable for schools in the U.K. but the restrictions of the school time-table make it difficult to give time to other than the technical problem itself.

The second aspect of technology in school is the question of devising an approach to the teaching of technology that will develop a pupil's imagination and ingenuity, and of finding ways in which these creative ideas can be expressed. I believe that these qualities are best encouraged by introducing key technical concepts in a qualitative and intuitive way, and that some of the most ingenious and creative work is achieved when a problem simultaneously excites a pupil's sense of form and aesthetic awareness, poses a technical problem and demands a practical realisation.

The Design Course at Pocklington School includes specific opportunities for technical creativity and inventiveness and this article describes the way in which these qualities are encouraged. The course is taken by all boys between the ages of 11+ and 13+ and has been fully described in references 1 and 2.

Encouraging Technical Ingenuity

The teaching strategies used in achieving this end must set out to widen and enrich a pupil's experience and to develop specific elements of craft and technical ability. Throughout our basic Design Course the aim is to give a strong impulse to original design by encouraging an individual response to a variety of problems. Where these problems are technical it is important that the underlying concepts are understood, intuitively if necessary, and that they are then used in a way that encourages individuality. However, ingenious and creative ideas do not occur in a vacuum but are a personal synthesis and association from elements of past experience; and the originality of these responses has much to do with the range and nature of the pupil's experience.[3,4] Thus this course, which has as one of its aims the fostering of technical ingenuity, gives a pupil a range of structured experiences of an artistic, craft and technical nature.

The ability to extemporise—to think in a divergent manner—is at the root of artistic expression and a similar mental excitement is needed to spark off ideas in the technical field. Through a study of natural and man-made forms, expressed in a variety of media such as pencil, pastel, clay, card, etc., a pupil is given an ever-increasing appreciation of visual values, form and structure. The excitement achieved in these studies, especially when they are carried out three-dimensionally, can provide the stimulus for suitable technical problems. This work in the art room constantly re-emphasises the basic nature of any designing, namely the need to experiment; for 'design, of every kind, is a matter of trial and error'.[5]

The generation of ideas can only be effective if the ability to express them receives equal attention. From the outset of the course a pupil learns to express his ideas rapidly through sketches, drawings or mock-ups as appropriate. The principles of orthographic projection are used when workshop drawings are needed, although the emphasis is always on conveying the necessary information quickly and accurately rather than on formal drawing. For example, in an exercise in which pupils conceive of, and then sketch, interpenetrating arrangements of rectangular shapes, the workshop drawing follows on naturally from the preliminary sketches.

In many situations a workshop drawing is inappropriate and a carefully constructed three-dimensional mock-up is the only way in which an individual's idea can be adequately expressed. The emphasis in the first mock-up is on presenting the basic idea for subsequent individual or group comment and rationalisation. The final mock-up will embody various refinements and the choice of modelling material will certainly influence the design. The importance of aesthetically pleasing shapes needs to be continually emphasised both at this and the subsequent workshop stage.

There are several ways in which a pupil's craft and technical experience can be developed; some situations will require specific directed craft experience through demonstration or class exercise where others can be solved by the ingenious use of standard craft practices. Technical concepts are introduced by structured exercises that concentrate on a qualitative understanding of key ideas with the minimum (or no) quantitative material. The third-form projects that arise from initial studies have involved a variety of lever and jointed mechanisms. The starting-point for all the mechanical arrangements was a series of directed experiments with simple card levers which rapidly led to the evolution of individual designs.

Another series of experiments with card, straws and thread served to illustrate the basic properties of tensile and compressive member forces within a structure. This work, in conjunction with the Project Technology slides on 'Bridges', provided the basis for bridge-building project similar to those described in reference 6. A similar approach is now being tried with simple gliders.

In conclusion, the imaginative use of technical skills demands a disciplined and ordered series of experiences on the part of the pupil. His imagination must be fed by direct observational drawing and followed by analytical studies which will deepen his intuitive awareness of colour, texture, structural and mechanical form. It is from these sources that his subsequent ideas will grow. He will need periods of craft instruction and the experience of working materials. Technical problems must be subjected to the same processes of investigation and analysis, and the underlying concepts treated in an intuitive and qualitative manner. It is vital that the work maintains its excitement at every stage, and the richest learning experiences will occur when the visual, aesthetic, craft and technical factors act as complementary parts of the total learning situation.

References

1. J. R. Jeffery, Art and design at Pocklington School. *Studies in Design Education and Craft*, **5**, no. 2 (Spring 1973), p. 29.

2. J. R. Jeffery, Design and creation at Pocklington School. *School Technology in Action*, p. 53, edited by A. Marshall, EUP and Open University, 1974.
3. Materials and design, a fresh approach. *Schools Council*, p. 96.
4. Children's growth through creative experience. *Schools Council*, p. 33.
5. D. Pye, The nature of design. *Studio Vista* (1964), p. 26.
6. *Simple Bridge Structures*, NCST, Heinemann.

IN A SIXTH FORM

MICHAEL MALONEY

Headmaster, Welbeck College, Worksop

Welbeck was founded over 20 years ago to help the Army recruit the technologically conscious and educated officers that would be required in the future. Its purpose is to provide the officers who keep the vehicles rolling, the communications buzzing and the computer display screens blinking. Yet at Welbeck and subsequently, they are not allowed to become too specialised since they are destined to play a wide variety of roles in technical management.

First and foremost, of course, they must be good soldiers and officers; hence the ability to work and lead a team. Secondly, they must be essentially practical; whatever discipline, they must be adept problem-solvers and makers of bricks without straw. These qualities during the 2 years of our sixth-form course are the Welbeck College hallmark.

And with five applicants to every place, we can be very selective: all are highly motivated and all are technologically alert. We have available to us the immense resources of the Army. Despite a very crowded curriculum we spend more time on subsidiary subjects and other activities than most schools.

A mandatory three-period-a-week course in the workshops teaches them the elements of engineering drawing, machine practice and so on. Over the last few years this course has widened considerably and we have developed the project approach and linked the technical department with physics, electronics, maths, computing, hobbies and art.

Our approach to project work is ruled more by practical considerations than educational ones. Thus a lot of the electronics is in no sense original but we still believe it provides a valuable exercise in problem-solving and analysis. The list below shows the range of projects recently attempted and arranges them according to whether they are essentially team activities or individual efforts. Some have been thought up by the boys, some they have been coaxed to do. Not all of them, of course, reach a successful conclusion and in 5 years' time our workshops will be a litter of projects which didn't make it! Not surprisingly, quite a number of them have some military interest.

Team and Group Projects

1. *Hovercraft*. This is led by a boy who was featured in the B.B.Cs 'Young Scientist of the Year' competition showing a radio controlled model hovercraft with a mine detector mounted on it. He has now moved on to the construction of a two-man hovercraft designed to be used as a rescue vessel on the College lake. Many design problems have had to be overcome during its development but the only difficult items to obtain have been old motor-cycle engines.

2. *Wind tunnel*. Plans are in hand for the construction of a wind tunnel for testing various devices.

3. *Wave tank*. This is an ambitious project inspired by work at Edinburgh University in which the conversion of wave energy into other forms will be investigated. It is planned to use electronic devices to monitor various functions and Science, Maths, Electronics and Workshops will all be involved. (See also *New Scientist*, 1 May 1975.)

4. *Satellite tracking*. An attempt to obtain pictures from the NOAA satellites which transmit pictures in real time as opposed to the ESSA series which transmitted a rapid TV picture from stored vidicon information. This was an early project which did not retain its momentum when the boys originally concerned left us. We did, however, obtain a rather crude picture of Scandinavia. It provided an example of the use of a programmable electronic calculator to solve design problems for the aerial.

5. *Closed-circuit TV camera.* This was a constructional project for the group taking the Schools Council 'Electronic Systems' 'A' level course. It involved some fairly advanced manufacturing and testing techniques and the finished model works remarkably well.

6. The same group have also attempted several of the Queen Mary's College sixth-form projects including the FET curve tracer and the optical communications link.

7. *Tele tennis.* A group of boys are working on the modification of a TV set to tele tennis but are finding it somewhat ambitious and are being held up by the difficulty of obtaining components. Another project of this type is an electronic piano.

8. *The ionosphere.* Various pieces of equipment have been built or modified with which it is hoped to carry out an investigation of the ionosphere by VHF reflection.

9. *Linear air track.* This project was given to a whole form who constructed a shuttle over a track about 8 feet long with relay switching. As the supply of air was provided by the Workshops vacuum cleaner it had to be dismantled after Speech Day.

Individual Projects

1. *Gyroscopes.* Inspired by Professor Laithwaite's Alice in Wonderland lectures. A number of simple gyroscopes have been made and their properties investigated in the workshops in conjunction with the Science Department.

2. *Telescope.* A boy is attempting to construct a 4-inch telescope from scratch and it is hoped that this will stimulate greater interest in astronomy in the school.

3. *Industrial archaeology.* A visit to the eighteenth-century scythe-making mill at Abbeydale stimulated two boys into an interest in this and they are now constructing a working scale model. This provides a link with the History Course provided for all boys at Welbeck.

4. *Antique weapons.* A number of boys enjoy the manufacture of model cannons and other replica weapons. The military connection is obvious but there is also interesting spin-off in the historical aspect. Primarily an individualist occupation and basically a test of craftsmanship.

5. *Biological amplifier* (Bio feedback). This was a most popular exhibit at Speech Day last year. It picked up brain waves and fed them back in a way that could be used for learning purposes. It is now being used to attempt to pick up electromagnetic radiation from plants.

6. *Radio-control devices for models.* Two fairly advanced systems are under construction and appear to be going well.

7. *Electronically-controlled position devices* have been built to direct an aerial for satellite tracking and also for a solar furnace which was constructed.

8. *An ultra-sonic transmitter and receiver.* An electronic unit for measuring ground resistance for archaeological investigations. Also a standard 'metal-detector' for treasure hunting!

9. *An investigation into paroptic vision.* Detecting light via the skin. Published in the College Magazine.

10. *An investigation into the effect of radioactivity on the germination and growth of cress seeds.*

11. *An investigation into the effect of noises on human powers of concentration.* (Another boy in the College was featured in the 'Young Scientist of the Year' competition for a project on the noise in discotheques.)

12. *Electronic devices.* This covers an enormous range, not only the usual hi-fi and stereo bits and pieces but also more outlandish devices referred to as 'fuzz boxes' and 'waa waa' units. Also model train controls, light dimmers, a set of electronic dice (!) and a timer for dark-room work. A light beam detector for the War Games Society to decide whether one tank can shoot another in 'rough country' is now being developed.

Three Case Histories

1. *The Story of 'Mavis'*

'Mavis' is a *M*obile *A*rmoured *V*ehicle *I*ndigo *S*ystem and in simpler language is a radio-controlled tank which will fire a rocket. The team of three boys who constructed it began with a radio-control device and the idea of using one channel to trigger a switch which would enable them to operate a number of different functions one after the other. They had heard stories of aeroplanes that came down and a submarine which would not come up so decided to concentrate on a land-based vehicle. The military interest in this project is obvious and it made a very good team exercise, each member of the trio having a clearly defined area of responsibility.

2. *Target Practice Simulator*

The idea is that a 'gun' when 'fired' emits a narrow collimated beam of light of low power but modulated at about 1000 Hertz. The target consists of a photo-diode connected to an amplifying circuit which is intended to sort out light modulated at a 1000 Hertz from the ambient illumination.

The laser offers the obvious and easy answer to this problem but he is trying to solve it without using a laser and there are signs that he might be able to get an acceptable beam width and response up to about 100 metres.

3. *Linear Motors*

This project was started some years ago inspired by Professor Laithwaite's book *Propulsion Without Wheels* (Pergamon Press). It got rather bogged down in the matter of the provision of a suitable three-phase supply and the orientation of the winding of the coils. In the last year, it has been possible to timetable a young physicist of a practical turn of mind into the workshops for about a quarter of his timetable to help with the selection and planning of the scientific side of these practical projects. The linear motors were the first that he helped the boys to sort out and several motors of varying sizes have now been made.

It soon became necessary to develop a longer test track than the original one and we now have approximately 30 metres of overhead track suspended above a side bench in the workshops. The erection of this track in itself was quite a project and considerable difficulty was experienced in putting up the 'trolleybus' wires for the supply of power to the motor.

The most successful motor weighs approximately 1 kg and runs at about 10 mph. The next phase of the project is to investigate the propulsion and at the moment an electronic toothcomb device is being developed to measure the speed and acceleration of the motor along its track.

About six boys have been involved during the last 18 months but it must be admitted that without the guidance of the Physics Master concerned, it is unlikely that they would have made a great deal of progress. Now that the thing is working satisfactorily and we have solved the problems concerned with its construction and the track, the boys are able to contribute more in the field of electronic monitoring of its performance.

Perhaps final mention should be made of the cost of project work. We are particularly lucky in having the resources of the Army behind us and although we have little money to spend we do have tremendous contact for obtaining 'bits and pieces'. My impression, however, is that, in the climate of educational stringency into which we are now moving, it will be essential for all project work to be designed around the facilities and equipment available. Our experience is that whenever we approach outsiders for help, advice or bits and pieces, the response is exceptionally favourable and encouraging.

WITH UNSTREAMED CLASSES

DONALD CHARLESWORTH

Deputy Headmaster, Norfolk School, Sheffield

The major aim was to explore the changes in the quality of life between the Stone Age and modern Sheffield. Incidentally, the children were to learn to observe, record, describe, analyse and criticise the world about them. The children were of S1 and S2 age, of widely varying ability in Rowlinson School, a mixed comprehensive, growing to its present 2000 population. Reading ages varied as much as 4 or 5 years even when the 'compensatory' children had been abstracted. Our classes were put together on a variety of bases, none of them academic; each class probably had two or three representatives from each of our feeder schools. Some found it decidedly difficult to string a sentence together, others were reading at considerable speed, by comparison, and working very quickly indeed. The stimulation of all the children was, perhaps, not as difficult as had been feared. The difficulty was staying in control of a situation which threatened to get out of hand, namely, the racers racing, the slow needing help. In the classroom, one fallacy that the bright child might help the not-so-bright was rapidly eliminated; expect little help from that direction. I am not convinced that the slower children get frustrated by being so far behind the very rapid ones. Much more apparent and difficult to deal with is the child who, at S1 stage, is already convinced that he cannot do something even before he has attempted it. Many children know they are not very good from sheer experience. Even so, they tend to underrate themselves and need massive encouragement.

Great care has to be taken to avoid a potentially more difficult situation, that of the bright child not being stretched to his limit. The temptation is to believe that he can 'work on his own'. Give him a text-book, a work sheet and he will be happy and successful and above all quiet. Unless he is in very frequent contact with the teacher, he will get frustrated and possibly very disruptive. He too should get his full share of attention.

Our departmental time-table was headed 'Humanities' and the time allocation was two 'doubles' per week, i.e. four periods out of thirty. The course was an integrated Geography/History/Social Studies course which was experimental and has now been through 6 years of development work. Field-work was a vital feature and a sympathetic time-table allowed it to happen.

The school site is such that I found it nearly ideal to stimulate real thought in children of this, or any other, age group. To the north is a south-eastern suburb of 1930s privately built semis; to the south a 10 year-old maturing Corporation estate of maisonettes, houses and 'high-rise' flats; eastwards the manorial village of Norton; westward, the 'village' of Greenhill—a microcosm of changing values over more than 300 years of English history. Sheffield city centre is but 6 miles away to the north-west. Each of the four sites mentioned is within 10-15 minutes walking time from school. Slightly further away, a stream through a public park gave opportunity for background work on the nature of rivers and their importance to man—leading later to a detailed study of water-power sites and the restoration at Abbeydale Hamlet, some 50 minutes away by public transport.

Almost all our field-work had to be within a double period of 1 hour and 20 minutes but a few welcomed sessions took half a day, the occasional one virtually a whole day, viz. Conisborough Castle and certainly York. Our first 'practical' out-doors was on the school field and within a few days of first meeting the class. The thirty-two children were each armed with a 50-inch plan of the school site and were asked to follow a few simple instructions and answer questions. They all completed the lesson knowing how to orient themselves and how to find north. There are few better ways of finding out the weaker and the stronger elements in a class. The writing up followed in the classroom confirming some of the ideas above. This type of activity was rapidly followed by plans of the classroom to reinforce our orienting lesson but also to confirm that all were 'happy' with the concept of scale; some apparently had never heard of it. The slower ones were given a basic drawing for which detail had to be measured and filled in. The quicker ones were able to do it all for themselves.

The syllabus was designed to include field-work, slides, films, tapes, descriptive writing, texts, abstracts, plans and maps. Last year, the first 2 years' course was restructured to culminate with the growth of Sheffield as a city, the second year's major theme being that of the 'opening up' of the world following the Renaissance. There were two dimensions of technology; that of the world at large and its effect on peoples and that of educational technology. We wanted children to see the effect on everyday life of inventions such as the plough, the water-wheel, the steam and internal combustion engines as well as high-rise flats. The other dimension was that we had to do this with virtually the total ability range and no extra money.

The team effort of making a plan of a hundred yards of stream bed through the local park was preceded by a lesson on how it was to be done. The children were paired off; one team of pairs was made responsible for the alignment of the base line's sighting staves and linking string, another team for taking equidistant offset lengths from the base line to the stream. The job was done in a 'double' period and easily completed during the next classroom session. The ability range only became obvious at the mapping stage but only as a matter of speed of performance. A good deal of 'helping' took place in the field but was relatively unobtrusive and apparently, therefore, more acceptable to the children.

The next aim with a 25-inch A 4 base-map was to survey Greenhill 'village' some three-quarters of a mile distant. There were two major purposes. The first was to survey building ages to establish the position of the older 'core' of the settlement, to map the building developments, to appreciate the way in which the settlement had grown and to attempt to see if, and why, its function had changed over the years. Secondly, a survey of building materials was completed and mapped, followed by a series of simple geology lessons which led to lessons on Stone Age man, his implements and way of life. The less able children were given base-maps with all the necessary buildings marked in and some extra help with the keys. Again, the more able children had to fill in their own.

Books ranging in difficulty from the Ladybird series to Longman's 'Then and There' series proved invaluable as well as quite a number of our departmental drawings of housing styles through the ages. The whole was 'tied together' by a discussion lesson using a series of my slides taken throughout the settlement. Children's comments about the work in the field are very revealing. During a survey of the Corporation estate came the comments, 'The gardens aren't big enough even for a game of Subbuteo' and from a not-so-able child, 'Why do they set all this grass round the flats and then put up notices saying 'No Ball Games'?' 'My grandma lives in one of those (high-rise flats) and hasn't seen her neighbour for a week'.

This type of work led to a study of Conisborough Castle, Abbeydale Hamlet and to a day in York. All were accompanied by very carefully thought-out work-sheets and had been preceded by several lessons to bring out the main features of the visits. The sheets allowed children to work at their own pace and gave structure to the work. However, they can tend to put a child off, if he sees too many of them: the skill lies in not making them too numerous, too easy or too difficult. There is also a strong case for varying the length and difficulty of those given to the not-so-able.

A stock of 'library' books from school, local and city semi-permanent loan resources are vital. The imaginative child will want to attempt to write an article on his own about 'My life as a knight in Conisborough Castle' or 'The railway comes to Sheffield'.

We hoped that our children had a good idea of the changing quality of life between the Stone-Age and the present and that they had some valuable ideas of their own to apply as they became more mature. Incidentally, they had absorbed a whole series of techniques which would be helpful to them.

Some Useful Resources

O.S. Map Sheets 1″ Sheet 111 Buxton and Matlock
2½″ SK 38 Sheffield
6″ SK 38 N.E.
SK 38 S.E.
SK 38 S.W.
SK 38 N.W.
¼″ Sheet No. 11

Royal Society: "Sheffield and its Region"
John Vickers: Series of Books on Buildings, etc., of Sheffield.
Raistrick: Articles in *Dalesman, Romans in Yorkshire.*
Department of the Environment: colour slides and information on Conisborough Castle.
Map, Guide and Notes on Avebury and Chysauster (personal colour slides).
Parish magazines: Norton Lees, Greenhill and Norton, Sheffield (personal colour slides).
Series of books from Longmans: "Then and There Series", e.g. *Norman Conquest, Prehistoric Man, Medieval Castles.*
Miller: "Water Mills of Sheffield".
Sheffield Museum's Booklet re. Abbeydale Hamlet Restoration.
Local History Series: Institute of Education: Sheffield University, Editor Mr. Batho.
Developing World, Longman.
Domesday Abstract for Hallamshire.
Life Magazine: Series on Stone Age Man.
"Humanities Dept." Rowlinson School holds many sheets of information derived from above and other texts.
Personal slides of Sheffield and Don Valley.
Film strips and colour slides—largely from "Visual Publications" and Longman (Common Ground) offering a very wide range of geographical and historical materials, e.g. "The Great Explorers" (U.P.), "The Australian Aborigines", Common Ground.

CO-OPERATION BETWEEN SCHOOLS

PHILIP A. COGGIN

Headmaster of the Park Senior High School, Swindon, Wilts.

The suggestion from an education officer that the building of the M4 across Wiltshire might provide matter for investigation was taken up by forty schools in 1967 who voluntarily collaborated, with varying degrees of effectiveness, over a period of 8 years. On the credit side was the stimulation provided by a challenging idea and the co-operation and exchange of ideas that developed among teachers of different schools, and particularly between the primary and secondary sectors. Although a book was produced on the first half of the project (*Birth of a Road*, Rupert Hart-Davis Educational, 1974), the secondary aim of producing a carefully documented study of the effect of the building of a motorway in its human and non-human environment was not achieved. Yet enough was achieved to demonstrate the potential of such a project as a contribution to sociological and environmental research. The help of schools in national land use and botanical surveys has been fully recognised and school children have also worked with local councils on traffic and other surveys. The M4 Project was essentially an extension of the classroom lesson but with professional direction and high-level support it could have provided hundreds of field workers over a large area for an indefinite length of time. Such a survey could be of great scientific value. For the children, professional guidance would have increased the authenticity of the work. Even so, involvement in a project that was outside and beyond the school heightened the motivation of pupils. As one boy put it: 'I am sure I voice the opinion of my colleagues when I say that I thoroughly enjoyed taking part in this project. From the fact that our piece of work will probably be of importance to many people, we derive a sense of achievement in completing it within a fortnight.' Official recognition of the work is therefore a desirable requisite.

Without this secondary aim as an objective, it would not have been possible to achieve the primary aim which was to provide an educational experience. The theme was sufficiently flexible to furnish a challenge for sixth-formers as well as opportunities for the primary school. With younger children, it was often possible to use the theme to integrate the whole curriculum: creative and descriptive English and the writing of reports; graphs, charts and percentages from traffic surveys; painting and model-making; elementary botany, zoology, geology, geography, local history and social studies.

At sixth-form level, studies in geography, biology and economics were often continued in depth at university or college of education and presented as theses for degree courses and teaching certificates. The eagerness shown by many of these students in spending week-ends and half-terms making ecological, land-use and farm boundary surveys was proof of the motivation that the project inspired. A memorable experience for six boys and girls from four different sixth forms was the presentation of papers at the Tenth International Youth Science Fortnight in London in 1968. It was fascinating to see how confidently they dealt with the very searching questions and comments that came from this high-powered young audience.

A great deal of the project work was specially prepared for exhibition in the classroom and for the whole school. There were also major public exhibitions in London, Bristol, Swindon and Chippenham. The preparation of work for display imposed its own discipline of accurate and orderly presentation. The standard reached by sixth-form scientists and geographers, with no pretension as artists or designers, was frequently highly professional.

No less distinguished was the work of younger pupils, for in one particular school, 300 fourth-formers were set to work, in twos and threes, during the post-examination period of the summer term, on a variety of traffic, social and village surveys. The plan was to spend 2 days on investigations and the remaining week of term on processing the results. Incessant rain during these 2 days severely curtailed the programme and proved a blessing in disguise because the information brought back from the few hours' work that was possible was so vast that the rest of the term proved totally inadequate for its processing. This has been a common experience in many schools though it is no doubt fully appreciated by professional researchers. A few hours spent collecting information outside the school will often provide material for classroom work for months. To witness the enthusiastic absorption of a group of children grappling with results of their investigation is to see real education in process. In fact one teacher complained to me that his big problem was getting his 15-year-old pupils to go home at the end of the day. Nor is this enthusiasm confined to the brighter pupils. The response of the less able is often spectacular.

The best teaching, it is said, is often done in digressions. Certainly, it is the unexpected spin-off that frequently provides the most exciting moments. The somewhat disruptive boy who arrived breathlessly at school one morning to say that he had asked his dad about his project: 'And do you know, sir, for the first time in my life he spoke to me as if I were a human being'. It was indeed a sobering thought that many pupils, including sixth-formers, returned from interviewing people in street and shop with similar reactions. 'They didn't treat us as school children; they spoke to us as if we were adults.' 'Every one was so kind.' And the greatest compliment of all: 'They thought we were from the Council!'

There were other surprises. A group of girls went off to investigate the effect of a new by-pass on a village and stumbled upon the case of a village that has completely disappeared within living memory. In addition to their original project they produced a fascinating paper on the depopulation of farming communities in Wiltshire at the turn of the century. A village primary school set out to compare the relative effects on their village of the old railway construction workers and the motorway workers. They ended up with a series of graphs, compiled from the parish registers, on death rates, incidence of epidemics, trades and occupations and the growth of literacy deduced from the ability of married couples and witnesses to sign the registers.

The experience of the 8 years of the M4 Project, compressed into a few hundred words, must sound more cohesive and spectacular than it was in reality. Yet despite its loose organisation and sporadic activity, it surely points the way to the possibility of developing the whole local community as a learning environment. A *Guardian* correspondent, Terence Benedixson, now a traffic consultant to the OECD, wrote a series of articles in 1970, on this project, in the first of which he said: 'This could be the single, most significant aspect of environmental education—the harnessing of many institutions to the task of explaining their work to children. Could it be that in the M4 Project there is a hint of a time when the juvenile ghettoes that we know today as schools will be split up and scattered throughout all adult institutions from building contractors to newspaper offices?'

Education for Survival

INDUSTRIAL STUDIES

PHILIP A. COGGIN

Headmaster, The Park Senior High School, Swindon

Like all the claimants for a place in the curriculum, Industrial Studies has a compelling argument: we live in an industrial society, most pupils will be employed on the productive, distributive or service side of industry, and all of them as consumers or voters will make decisions with industrial and commercial implications; it is logical that they should have a general background of the industrial scene. The logic does not, unhappily, create space on the curriculum. Like other 'new' subjects, however, Industrial Studies often looks at familiar material from a different aspect or new information through traditional disciplines. Rural Studies, for instance, acquired a new dimension under the influence of the environmentalists and the interaction between town and country inevitably led to the idea of urban studies and urban trails. The place of industry in all this is obviously crucial but it is less a matter of imparting information, and thus breaking the back of an already overloaded curriculum, and much more a question of opening windows and eyes and creating new attitudes.

In the field of careers, the new emphasis is on education rather than guidance and much of the modern syllabus is designed to make explicit what has previously been taken somewhat for granted. Its aims—self-assessment, decision-making, social awareness, and a knowledge of the world of work—are clearly very relevant to industrial studies. In all disciplines, new and more relevant subject-matter could probably, with advantage, replace less effective material. We could learn much from the past and from other countries. In 1658, for instance, Comenius published his *Orbis Pictus*, a small illustrated textbook for teaching Latin through the medium of a comprehensive outline of contemporary society including descriptions of all major trades, industries and other social institutions.

In Swindon, two conferences of employers, trade unionists and teachers, together with extensive surveys, carried out by teachers, of the requirements of employers, have revealed a considerable consensus of opinion about what the schools should be doing about preparing pupils for the world of work. Teachers, on the whole, felt that these views also strengthen their hand on purely educational grounds.

Employers spelt out their ideas in the plain, practical terms of the office or factory floor. If a single statement were needed to summarise what they said, it could take the form of the Georges Gusdorf quotation which heads the chapter on Language and Learning in the Bullock Report: 'Man interposes a network of words between the world and himself, and thereby becomes master of the world.'[(1)] No employer mentioned the Bullock Report, but by far the largest single topic was the question of communication skills.

In even the most lowly tasks, the ability to talk, read and listen to instructions, to make reports and record work done, accurately and clearly, are of crucial importance. 'So much frustration,' said one employer, 'is caused in industry by the inability to convey one's thoughts accurately.'

Communication and co-operation are complementary. The whole process of modern industry depends on team-work and there was a universal plea that the schools should encourage pupils to work together in groups. One teacher summed up the general opinion by saying: 'It is obvious that one unco-operative, uninterested or selfish operative can destroy the harmonious relationship essential to this work structure.' There are deep implications here for the development of character and intellect. The essential ingredients of team-work, identified by many employers, were motivation, co-operation, communication skills, sociability and getting on with people and, high on the list, good manners, consideration, punctuality and attendance.

Numeracy came in for less comment than literacy and oralcy but nevertheless bulked large. Modern Maths was criticised and many teachers were surprised at the extent to which traditional weights and measures, and therefore fractions, are still used as well as metrication.

The key to industry's growing concern about educational standards seems to lie in the question of motivation which, in many recruits, is conspicuous by its absence. Some firms drew attention to the importance of creative forms of leisure as a motivation for work, especially boring work. Employees often worked hard in order to be able to play hard. Something to look forward to, other than television or alcohol, puts work in perspective. If leisure, in Whitehead's phrase. becomes the 'seed-time of degeneration', work suffers. Even overtime can, in the long run, be destructive by becoming a substitute for nothing else to do.

One solution to the problem of motivation would seem to lie in approximating the world of school more closely to the world of work. There appears to be no basic conflict between the views of industry and those of the school. If teachers felt a slight suspicion that demands for obedience to instructions and respect for authority were a veiled request for factory fodder, it was soon realised that employers were not talking about authoritarianism. On the contrary, they emphasised the need for initiative, self-discipline, responsibility, resourcefulness, adaptability, flexibility, originality, self-confidence, mature independence and the ability to think for oneself and work on one's own without supervision. 'We prefer quality to qualifications', said one employer, voicing a general opinion that academic examination results were useful only as indicators of a general level of ability or as necessary to meet the requirements of technical colleges, polytechnics or universities to which the employee might be sent. Intrinsically they were largely irrelevant except as evidence of a candidate's ability to complete a course.

Having forcefully enunciated general principles, employers were reluctant to comment on the subject-matter of the curriculum. A sizeable majority, however, would welcome more emphasis on current affairs, on the history and function of local firms, on a balanced non-political assessment of the roles of management and trade unions, and on basic economics and civics. 'The Facts of Life in their broadest sense,' as one employer put it.

These views direct from the grass roots and gathered by forty teachers from as many different firms, echoed with extraordinary fidelity the official line of the CBI. In their evidence to the Schools Council Working Party on the Whole Curriculum,(2) the CBI expressed their anxiety about early specialisation on the one hand and on the other, their belief 'that liberal examination and curricula trends ought not to be allowed, in the interests of the pupils themselves, to lead to a dilution of traditional standards of competence in areas like numeracy and literacy'. Equally valued is involvement in the traditional activities of a British School: 'Experience has shown that employees who have taken a full and responsible part in school life—games, social activities, clubs and societies—or who have been helped to develop their own leisure-time interests, are generally more successful in their jobs than those who have not'.

If this sounds like a plea for a return to the good old days of the Three R's and 'Up the School', it is because employers are adamant that the fundamental aim of a good, all-round general education up to the age of 16, should be preserved at all costs. The irony is that educationists accept this in theory but not in practice and, as the Crowther Report[(3)] put it, 'they refuse to will the means, and therefore must be held to deny the end'.

Industry appears to be appealing not only to a much older tradition of a broad education but is also asking the school to take much more literally the traditional aim of 'Education for Life'. Their plea is therefore more revolutionary than it sounds. This runs throughout the CBI's evidence: in their call for greater efforts on the part of the schools 'to instil into pupils a clearer sense of purpose and direction'; in their concern that 'some teachers may be creating an unfair picture in the classroom of the employer as a "phantom tyrant" and that "uniformed discussion" might create "an anti-industry atmosphere"'; in their belief that 'teachers ought to be encouraged to involve themselves with local industry and commerce in as many ways as possible'; in their view that 'basic science and technology must form part of the curriculum even for those wishing to specialise in art subjects'; in their hope that the new 16+ examinations will 'result in a more broadly based test, reflecting the candidate's performance over the whole course and not merely a few hours on the examination day'; in their insistence that 'vocational education (as opposed to training) . . . is very much part of a 'general education for life' . . . covering the many aspects of working and domestic life—taxation, voting, life assurance, mortgages, trade unions, etc., and that during the last year of compulsory schooling, the activities in which pupils 'might be involved away from the traditional atmosphere of school, though still under its jurisdiction and as part of its organised curriculum, would be essentially outward-looking'. All this points to the desirability of combining the traditional general education, not with specific courses of a vocational character, but through 'a change of emphasis in certain existing subjects' with the need for teachers 'to relate school studies to the personal and working lives of their pupils'.

Many teachers were able to discuss industrial relations with both sides of industry and there were two local conferences on the subject. The rest of this article is my own summary of the views expressed by trade unionists.

The image of trade unionism among schoolchildren is formed almost entirely by the mass media. With few exceptions pupils regard trade unionists as selfish, militant, unpatriotic men whose one aim in life is to stop willing workers from getting on with their job. In the summer of 1974 the TUC decided to take active measures to dispel these prejudices and promote an objective view of the movement. They approached the Schools Council with the suggestion that there should be a major national curriculum project on Industrial Relations. There was an immediate outcry in the press that the TUC were trying to obtain control of the school curriculum. This was quite untrue. On the contrary, they were willing and anxious to expose their case to objective and critical examination so long as they also had the chance to put it in the first place. They welcomed the desire of the CBI to be involved as a partner in the project.

Teachers are wrong to believe that what goes on in the schools is the preserve of the teaching profession. All sections of the community have a right to hold a view about the school curriculum and the TUC have a view which ought to be heard. They have had a view on the curriculum since 1879 when they wrote to the Archbishop of Canterbury complaining about the condemnation of trade unions contained in the textbooks of church schools.

Now that the TUC wishes to become actively involved in the debate about the curriculum, it is pertinent to ask what claim Industrial Relations has for inclusion rather than some other subject. The case rests on the present imbalance of the curriculum. Great attention is rightly paid to the development of pupils as individuals. Sport, music, drama, literature, art and other subjects all provide children with opportunities that will help them to take their place in society as individuals and to pursue individual interests and leisure-time activities. Much less attention is paid to man as a political and social animal. The collective ethic, which includes industrial relations, politics and law, is as important as the primacy of the individual.

If British democracy survives, schools, because of their neglect of the collective ethic, can take no credit. Pupils are leaving school politically and industrially illiterate and are left to pick up ideas and information haphazardly from inaccurate or biased sources. They cannot take part effectively in the decision-making processes of society.

Now that the divine right of employers has gone, are we training people in the right way? Schools are about the development of children so that they can go out suitably equipped into the world. It is not a question of vocational training but a demand for a curriculum that is relevant. Since all children will take their place in society as future workers, it is surely essential that they should start with a basic knowledge of the vocabulary and facts of Industrial Relations.

If Industrial Relations are to be taught, then we are forced into considering economics and politics. Teachers, however, are not trained for this and there is an almost total absence of teaching material on the subject. This must be remedied and help obtained from the practitioners. The challenge is an ethical one. The most that can be said of the present curriculum is that it is innocuous but teachers should not be neutral when faced with a whole new world of value judgments.

The key themes in Industrial Relations are:

1. The evolution of Industrial Society.
2. The specialisation of labour.
3. The role of capital.
4. The growth of labour organisations and the status of the employee.
5. The social and moral values that underlie industrial society.
6. The role of management. Who gave the employer the right to dispose of resources?
7. International monetary affairs.
8. The Social Wage, e.g. contrast the system of Japan with that of Great Britain.

9. The language of the Industrial Society.
10. Strikes, procedures for bargaining, voting, disputes, etc.

Industrial Relations in schools can be taught through simulation and role-playing techniques. Pupils can be placed in a situation and asked how they would respond. Case histories of local disputes and how they were resolved would supply suitable background material, managers and trade unionists being invited to take part in discussions. This is not a plea for indoctrination or docility but a chance for the rising generation to do better than us. It is a plea to the schools to play their part in the regeneration of society.

References

1. *A Language for Life*. Report of the Committee of Enquiry appointed by the Secretary of State for Education and Science under the Chairmanship of Sir Alan Bullock, p. 47, HMSO, 1975.
2. *Supplement to Confederation of British Industry Training Bulletin*, August 1973.
3. *15 to 18. Report of the Central Advisory Council for Education—England*, vol. I, p. 269, HMSO, 1959.

SAFETY . . . ?

PHILIP A. COGGIN

Headmaster, The Park Senior High School, Swindon

> We all live under strain, yet the adrenal cortex thrives on it, and total freedom from stress is death. How can we best learn to live with stress?
>
> Prof. Hans Selje in the Magazine of the World Health Organization, Dec. 1974

How has it also come about that a technological society which has vastly increased the risk of death, injury and disease, in almost every sphere of life, yet manages to effect a constant overall improvement in the health and life span of its members?

These two questions, one in the epigraph, the other in the opening paragraph, must have occurred to any reflective teacher in advanced Western society, and must have prompted him or her to wonder whether the pupil's capacity and motivation for learning could be expanded by extending the learning situations. Sensitivity to the variety of experience and imaginative response to the data perceived are the much neglected but essential foundations to any disciplined approach to education. Accidents, almost by definition, are unique occurrences and no amount of instruction can cover all eventualities. Technological society requires the questioning mind that gives answers to the constant supposition: What would happen if . . .?

The questioning mind can learn throughout its waking hours from home, street, part-time job or, most of all, from its fellows since children communicate much along themselves. If thirty pupils have received the same piece of instruction, what is left for them to talk about? But ten trios (divide them how you will), engaged on many different enquiries, are likely to have something to communicate and to discuss. Many experiences testify to the fact that when a joint project is sparked off, there is created an intellectual climate which the pupils carry round with them outside as well as inside the classroom. Thus a sixth-form by manning (womaning would be more accurate) a crossing for a local primary school has begun to discuss and argue about the speed and discourtesy of some motorists.

So too a survey on smoking habits by a group of 15-year-olds has led to some lively discussions about the dangers and the expense of cigarettes. It is possible too that those interviewed gave more accurate information to the fifth-formers than they might have given to professional sociologists. Self-generated discussions of this kind are likely to do more good than straight talks or horror films on lung cancer.

One group of 15-year-olds who found writing rather difficult, went out with a camera and a portable tape recorder. In a matter of hours they recorded thirty-six incidents illustrating infringements of the Highway Code. The transposing of the tapes into writing for exhibition purposes was a further excellent exercise in English. The designing of Safety posters is already a well-tried method of stimulating interest and setting off the imagination. It also gives the art department the opportunity of involving itself with other subjects. In a more specialised area, a fifth-form boy became interested in the effect of drugs on driving. Such a project was necessarily derivative, since practical experiments were obviously impossible. He ferretted out the literature, reviewed and classified the whole range of medical and other drugs that might be taken by drivers and discussed the effects they produce. The local Road Safety Committee were so impressed that they submitted it to the Royal Society for the Prevention of Accidents who had it printed and circulated. The boy received back a copy heavily annotated by the medical consultant to ROSPA, who treated it as seriously as if it had been an essay from a medical student. The amount of discussion among fellow pupils to which this investigation gave rise is again likely to have been more productive than a straightforward lecture on drugs.

The building of a motorway through the neighbourhood has led another group of pupils to investigate the safety of motorways. From information obtained from the Road Research Laboratory they made graphs, comparing the accident rates of an area before and after the building of a motorway. Locally, the AA and RAC were interviewed and figures obtained about the incidence and causes of breakdowns and accidents as well as their views about bad driving habits. A request to interview the motorway police was met by an offer to give an escorted tour of the motorway by coach. Accordingly, forty pupils accompanied by three police officers travelled along the motorway to a running commentary from the experts. They visited the maintenance depot and learnt about the work of the county motorway surveyor. Among other things, he discussed birds and animals as traffic hazards. The fact that he had kept records of the dead birds and animals retrieved from the motorway led one group of pupils to follow up this line from the biological angle. Were some species, for instance, learning to live with the motorway? The coachload of pupils went on to visit the police depot and the service station, to learn how accidents were dealt with, and to receive and discuss the figures for accidents and breakdowns and their causes. The final visit was to the control room at Police Headquarters in Devizes from which the Wiltshire section of the M4 is supervised. The forty accounts of this expedition that were written by pupils of all abilities were remarkable for their length, enthusiasm and grasp of detail as well as for the variety of emphasis and impression.

The interest aroused by this particular project led an aspiring medical student to be admitted as an observer at the casualty receiving station at the local hospital to see, at first hand, the results of road accidents. Another group were directly involved with the local Council in making a seat-belt survey at the town's car parks before and after a publicity campaign.

The AA and RAC have co-operated with driving schools in schemes to give senior pupils instruction in driving and in car and motor-cycle maintenance. The community at large would certainly feel the benefit if this most vulnerable and potentially dangerous group of young people all left school as better drivers and more socially conscious road users.

These examples, covering only a few areas of this extensive field, point the way to projects in other aspects of Health and Safety education. How, for instance, might the Biology and PE departments collaborate in monitoring weight, pulse rates and other reactions in a variety of situations? How might the Home Economics Department tackle the problem of that most dangerous of all threats to our health, too much of the wrong kind of food, and the problem of that most dangerous of all accident hazards, the Kitchen?

ANTICIPATING THE FUTURE

HAROLD CRAIG

Headmaster, Commonweal Senior High School, Swindon

There are innumerable alternatives to the ways in which we can outline the ultimate aims of education and examples can be found in journals and elsewhere. I recommend that we change our point of view slightly and, before specifying what we think might be the best treatment for the developing pupil, we examine the roles which he will eventually take on and which, I suggest, are the four listed below. They are not given in order of importance because the order will vary from one person to another and from time to time in the same individual depending upon his occupation and interests at any moment. All pupils will eventually become:

(a) members of a society whose way of life has developed along technological lines without much regard to the ultimate consequences of the present trend;

(b) members of a community whose history and background have for centuries been one of rule by democratic methods;
(c) a unit in an economic structure in which the welfare of the whole is dependent upon its members putting in at least as much as they take out;
(d) individuals in their own rights, each having certain characteristics in common with their neighbours but also having those aptitudes, preferences and attitudes which make them unique.

I am not advancing a revolution in educational method or curriculum which will provide the perfect answer, but I do suggest that we look at the four roles in turn and consider how we can best prepare our students for them.

(a) The ingredients of that part of teaching designed to take care of technology should not only be the general scientific principles directed towards modern uses of resources, but also the study of the limitations of those resources. We are beginning to make a fair job of the former in some respects. Jet propulsion, solar energy, nuclear energy, satellites and the like are in science syllabuses. Where we are falling woefully short is in the teaching of the difficulties which will face us in future years by the reckless use of what we have available to us, and those years could well be in the lifetime of the very students now passing through our hands. Our chemical natural resources, gas and oil amongst them, are strictly limited. Although the use of nuclear energy is extended from uranium to plutonium, there is no way in which we can build up supplies of useful radioactive chemicals for nuclear energy. We must content ourselves with and make the best use of amounts already present in the earth. In any case, pollution is already a major problem with no signs of a solution. The alternative sources of energy available are all due to the Sun or Moon, all virtually without limit, and except for one or two mountainous countries with a good supply of water, hardly utilised. It is now that these sources should be tapped and the Earth's resources stored for carefully controlled dissipation over the thousands of years for which they will be required. So our up-and-coming students should be made to realize how essential this aspect of our future has become.

(b) My second point was democracy. If we are to remain as we have been for centuriés, a country governed by the people in free unfettered elections, it seems to me that not only civic or social studies are essential but that they should be extended to include an appreciation of the need for an organisation, whether in factory, office or school, to be a unit which can only work for its own ultimate good if it does so in harmony without the disruptive operations by individuals or minorities. The warning signs are already up for us all to read.

(c) My third responsibility of the adult follows closely on this one, i.e. that everyone is a unit in the economic structure of the country. Is it here that we are falling down most seriously in industry? Taken overall during recent years, there seems to have been a lack of understanding of the importance of a balance of payments and other economic realities, a selfishness born either of ignorance or of disregard for the consequences. A course of basic economics might go some way towards taking care of this problem, and there should be as much contact with the post-school situation as can be arranged—work experience supported by down-to-earth but reasoned and knowledgeable talks by trade-union officials, managers and the like. A good deal of self-discipline and moral judgement is essential if those in strong positions are not to bare their fangs against public opinion.

(d) an individual in his own right. Much greater attention has been given to this during the past 10 or so years and better opportunities have been offered for work and leisure to enable students to find out what they enjoy doing and what they are good at.

What then would I do with the curriculum? It is not so many years since students in secondary schools had a basic course of what were then considered to be the essentials and in the fourth year a small range of options which they carried for 1 year if they left at 15 or 2 years if they stayed at school to 16.

Over the years since the introduction of the GCE in place of the School Certificate—and how our lives in schools are influenced by exams:—there has been an increasing tendency to offer more subjects as options at the end of the third year until now I have counted forty-four subjects listed for 'O' Level in our Examining Board, some of them with several alternatives—and a few more for CSE not present in 'O' level. And this total does not include a wide range of special languages and the various alternatives available under many of the subject headings.

I would go back to the situation prevailing some years ago—not in subject matter but in restricted range. Being more specific I would have the basic Maths, English, PE/cultural group supported by the extended General Science/Technology to which I have referred, embracing Conservation or Environmental Studies or whatever you like to call it, and supported also by the modified Civics-cum-Work-Study. This leaves time for some options, the discussion of which is beyond the scope of this article.

I realise that the effect on the syllabus would be quite marked and that the level of entry to specialist courses in Further and Higher Education would be lower so that for the same duration of course the standard at the end, e.g. of a degree or diploma course, would be lower. North America has shown us that this is no bar to their research students reaching the highest level. The benefits to society and industry in attitude could well be appreciable. Certainly those families which are required to move from one area to another through employment (more of them every year) would find their children fitting more easily into changes of school. The benefits to the schools in improved efficiency and simplicity of organisation, better use of staff and equipment, lower overall cost, are too obvious to dwell upon.

SURVIVING THE FUTURE

VINCENT RAWLINSON

Electrical Engineer, National Youth Officer,
National Voluntary Civil Aid Service

Preparing for an emergency that may never happen is not an objective calculated to inspire young people with enthusiasm. The threat of an atomic attack does not create among them the awe that we of the last generation feel. For them it is just a technological development of today and to some degree accepted. Yet disaster from nuclear war is no less possible than disaster from pollution or the exhaustion of energy resources.

Having met qualified instructors from the Civil Defence, I strongly believed in their value to the community in preparing people to be useful and ready in any emergency or major disaster. I therefore offered my services as training officer and soon found that a new image, that of Civil Aid, must be developed and designed to suit the requirements of the particular community in question, and that younger members must be introduced into the unit. As a former Scout leader, I saw the possibility of a movement that boys and girls could join not only for excitement and involvement when young but in which they could also progress to adult membership helping society and becoming an effective part of the community in which they live.

We found that the best age groups to work with were 9 to 11-year-olds and 12- to 16-year-olds. We then devised a training system based on Civil Defence and some Scouting skills. These groups took part in all the civic parades and became as much involved in the community as possible. The training was put into operation during the winter once a month. Our aim was to produce a group of young adults (cadet officers) of 16 years of age or over, qualified in adult subjects to form teams of active Civil Aid members helping in the community and being prepared and ready for a major disaster of any kind. In a broader sense, we hoped also that as they grew older they would become more responsible members of their community, useful to others as well as themselves.

We found that at 11 years of age and over they were prepared not only to take adult courses but, through their NCO meetings, they insisted that by the age of 16 a member should have taken three adult courses to qualify for the young adult section. These might be Nursing, First Aid, Wireless Operation, or Rescue Training. The young adult sections operated like Venture Scouts or Ranger Guides but were part of the Adult Civil Aid Force.

A most important development was the discovery of the crucial importance of showing the members of the 9 to 15 year age group that their individual talents, thoughts, ideas and problems were appreciated, that they were all listened to with understanding, and that the leader, in the eyes of the young people, gave real assistance in working out their problems as well as their ideas.

We instructed the 14-year-old boys and girls to be radio operators but it was of no avail to sit them at a radio with a microphone in the hand, and attempt to teach radio procedure. The machine soon lost its appeal.

So we arranged a group in a circle using only things they themselves understood. They played a game called 'Matthew Mark'. Each child calls his or her name or number across to another name or number in the circle. The next turn is taken up by that member saying his or her name or number and calling another. Any wrong procedure or hesitation and he or she moves to the bottom position. After playing for 15 minutes the names and numbers were changed to the phonetic alphabet, Alpha, Bravo, Charlie, Delta, etc. They never tired of this game and all the 9- to 16-year-olds played it happily together. In less than 3 weeks, even the 9-year-olds knew the phonetic alphabet. We then instructed the 14-year-olds in simple radio procedure and, after a further 3 weeks. gave them the job of using radio in an important local exercise. Five boys and three girls took part and were declared by the long-qualified instructors of the Civil Defence to be superior to the average adult operator. They were more reliable under pressure and their keenness and interest would not permit slipshod or wrong procedure.

Children can come to terms with the machines of today but in their own way and at their own level. To interest young people we have to find the correct basic equipment, enough to be of interest but at a level that will make the individual feel master of the situation. Then they can advance to more sophisticated situations. This was true of all activities, whether in camp or in the lecture room.

The crucial time came when the youth attained a standard in their special field equal to, or better in some subjects at the age of 16 than the adults. It is unfortunate that when these young people have reached a point when they are equipped with all the basic knowledge and all they need is experience, they are often put down or rejected by the older and by no means superior teams who can only remember them as the children to whom they gave parties and pointed out their faults during training. At the 1975 Youth Conference held at Birmingham Civil Aid Headquarters we found that Civil Aid youth in different parts of the country, in different circumstances and with different community needs, had undergone the same experience and it was found that membership included all social classes and standards of education.

The young people have become involved in the community by volunteering to help in many ways. These have included deaf and dumb language, help with animals, assistance with spina bifida children and with older people, visiting hospitals, assisting at carnivals, fairs and sporting events. Any achievement by young people, whether working for themselves or for other people, should be acknowledged as soon as possible after the event. Too much publicity is given to the anti-social elements and there is a danger that they may become the heroes of the day.

Finally, I have found that a youth leader must be prepared not only to listen to the problems of an individual member but must be in touch with the family, school teacher, priest and any other relevant contact. It is most difficult to assist young people of today in their leisure pursuits if a start is not made before 15. If we can reach the 9-year-olds and upwards, then the 16-year-olds develop respect and, provided they are accepted by the older adults, will work to help the society of which they are part and probably become better men and women than those who gave them instruction. This is what I hope for the future.

The Way Ahead

TECHNOLOGICAL LITERACY

EDWARD SEMPER

National Centre for School Technology, Trent Polytechnic; Formerly Headmaster, Danum Grammar School

Why is it essential that we should educate all our young people in ways which arouse a delicacy of feeling for technology and develop a more positive stance towards it?

In the first place, because our livelihood is at stake whilst British industry continues to decline. 'The next greatest disaster to defeat is victory': Wellington's prophecy was never more true than in the first flush of the British industrial revolution. The stream of inventions which poured forth from this country in the late eighteenth and early nineteenth centuries were largely the achievements of self-taught men. They sent their sons to public schools where the curriculum bore no relation to technology either as a cultural influence or as a means of innovation. Unwittingly, Dr. Arnold fostered a disdain for the practical arts among young gentlemen, and technology became the province of men with practical ingenuity who cherished a contempt for education and developed unreasonable attitudes to change. Fierce competition and unbridled exploitation led to the first hardening of attitudes between management and men; no longer feeling to belong to a single fabric of society, men sought security within the more familiar boundaries of new social classes. Patterns of behaviour in both industry and education which formed in these early days seem now to have become part of our national psychology and, although currently manifest in different ways, these still threaten our future prosperity.

In the second place, because the extremely rapid advance of science and technology over the last 50 years brought new tensions into society. Instant communication of world-wide events, self-service at petrol stations, computer-controlled car parking, pills, tranquillizers, and even Moon landings are now taken for granted. Yet, in recent years, there has been, as Bronowski puts it, 'a retreat from knowledge'—a withdrawal from science and technology—and a dabbling with mystery and the occult. Could it be that in the process of advance there has been such an erosion of cherished beliefs that we are now witnessing the first effects of a loss of nerve? Is it not the case that without a belief-system every man is free to do what is right in his own eyes: to plunder or to conserve, to be promiscuous or to be faithful, to violate or to enrich the quality of life? And is there not some evidence of an intuitive fear that we may be on the verge of even greater change over which we might have little control? Professors Toffler, Bell and Stonier consider that already the industrial society is being overtaken by the communicative society, foreseen by them as a service economy increasingly dependent upon computers, television, video-telephones and new electronic devices, and upon automated organisation and production. Shall we, as they believe, find ourselves increasingly selling skills and know-how in providing services—new systems for designing and planning the complex wiring which goes into power stations, ships and North Sea oil rigs; new machines, appliances and processes to achieve more with less consumption of energy and materials; new strategies and improved services to tackle a host of social, environmental and educational problems, for example—and ought we now to aim to become the world's 'think shop' rather than 'work shop'?

Whatever the speed of transformation, it is apparent already that transition to a post-industrial society will be immensely difficult and that if severe disruption is to be avoided, we shall need a very much changed attitude to technology. People need to feel in their bones that it is right to conquer and to develop the universe. They need to know that 'having more' can really mean 'being more', that technology can have 'ideals as well as ideas', and that it can be used to accomplish an unprecedented development of the earth's resources for the benefit of all mankind.

This then is the *raison-d'être* of universal education in technological literacy. It is wholly consistent with the plea of the two Archbishops and with the ideals of comprehensive education. In a stimulating article 'Political ping-pong in our schools' (*Sunday Observer*, 16 October 1975) John Day pin-points the two essential tasks of comprehensive education which are widely acceptable amongst headteachers and yet relate closely to Dr de Bono's view of teaching technology in schools. The first task is to recognise that dignity and a sense of worth are owed to all human beings irrespective of talents, social status or income. It is impossible to discharge this responsibility unless one deals with 'the nature of technology, its possibilities and its limitations, the dangers and the impact of technology on society' which de Bono recognises as part of everyone's general education. Dignity and a sense of worth are denied to those not 'equipped to live in a technological age', 'unable to communicate with the environment in which one is going to live'. The second task is to recognise and develop the differing degrees and kinds of ability that children possess—a process which inevitably leads, sooner or later, to specialisation of some kind or another. This corresponds to de Bono's second aspect (or subject) of technology in schools, viz. 'design, invention, problem-solving, testing, cost-effectiveness, how to bring about an effect and avoid another and the like'.

There is here a clear distinction between how schools should respond to the needs of all children as potential citizens of a technological society, and how they should meet the needs of young people who feel committed to or are interested in specific aspects of technology. Yet these tasks are never wholly separate: when one talks of developing technological sensibility—essentially a gradual process extending over the full span of school education and beyond—one appreciates that this can never be achieved without first enabling young people to experience for themselves something of the inner structure of technology, and of what is meant by an inventive turn of mind and a technological approach to problem-solving. Children need this kind of experience in order to begin to appreciate in what ways, for what purposes and by what means technological development might usefully be stimulated and controlled.

What more should be done? Firstly, and perhaps most importantly, we ought urgently to seek ways and means of developing a delicacy of feeling towards technology amongst very young children (5 to 9). Secondly, we should extend opportunity in the middle years (9 to 13) for imaginative and inventive thinking both as means of identifying aptitudes and abilities, and of developing universal technological sensibility. Thirdly, in the upper school we should develop thematic programmes of action-based learning which relate technological innovations and developments to the economic, social and cultural changes which flow from them. This implies the provision of a new element in general education, viz. technological literacy. It is based on

- observing and assessing the precise nature of technological achievements and their potential for development;
- exercising imagination and inventive thinking, and employing decision-making techniques and problem-solving skills to relate technological innovation and design to social, scientific, moral, economic and political criteria;
- anticipating trends in technological innovation and envisaging ranges of possible consequences;
- managing genuine work activities within the school environment (with break points for sharpening skills by the use of role-playing and the solution of case-study problems);
- making realistic self-assessments of the work-roles and life-styles of technologists and technicians.

Action-oriented education of this nature provides a meeting ground for teachers of all school disciplines with a concern to find new ways of responding to genuine need in this age of advanced technology. As a mode of entry to the curricula of all children, it would doubtless be welcomed by enthusiasts for political education, careers education, European studies, environmental education, technological and design education, and education for leisure and in industrial relations.

In the next article we consider how best to equip rising generations with the knowledge, skills and attitudes to participate effectively in open debate of alternatives in technological change and to contribute to their development.

STRATEGIES FOR DEVELOPMENT

PHILIP A. COGGIN

Headmaster, The Park Senior High School, Swindon

and

EDWARD SEMPER

National Centre for School Technology, Trent Polytechnic

> The life of man is founded on technology, science, art and religion. All four are interconnected and issue from his total mentality. (A. N. Whitehead).

What questions should we now be asking in order to decide how best to achieve a more positive stance and constructive attitude toward the technological society in which we live? Perhaps the most important are these:

(i) What sorts of experiences are needed to develop technological sensibility at an early age and to forestall unconscious prejudice?

(ii) What kinds of knowledge, attitudes, values and skills enable young people to begin to comprehend the economic, social and cultural changes which flow from technological innovation, and to become involved in ordering priorities and choosing between alternatives?

(iii) In what ways can these be presented as valid educational experiences to children of widely different backgrounds and abilities?

(iv) What is to be the relationship of education in technological literacy and sensibility to the whole curriculum?

(v) How might these changes best be accomplished?

Education in technological sensibility ought to start in the primary school for moral attitudes are determined early in life. Fun, fantasy and hero-worship should be the hallmarks of inventive thinking which can be stimulated by dramatic presentations of basic concepts on which our technological civilisation is founded and of the lives and work of heroes of technology. Dr. de Bono's thinking course for juniors 5 to 12 years, the work of Sir Bernard Miles at the Mermaid Theatre, and Kevin Walton's lively presentations of case-histories of technological innovations suggest ways in which the work might be developed. At the earlier ages of 5 to 7 years, fantasy and myth are not wholly separate from reality in the young child's viewpoint and this offers an added opportunity for teachers appropriately to personalise and vividly to illustrate the life and work of great inventors, discoverers and engineers.

In junior and middle schools, the history of early technology offers plenty of opportunity to observe, draw, design and make, using the simplest of materials. Ships, water-wheels, windmills, clocks, looms and calculators are not less attractive because they offer less sophisticated technology. Junior and middle school children are fascinated by things which move and work: they love to make them work better. At this stage it is essential not to stop the free flow of fantasy by undue haste in the introduction of analysis and quantitative evaluation. The work of R. N. Billington and J. R. Jeffery at Pocklington School is outstanding for its emphasis upon observing natural forms, free speculation of related technology, and design and production based on qualitative and intuitive understanding. There is here a remarkable parallel with the sequential approach of G. J. Fox and D. F. Marshall to the child's intuitive understanding of the control of automatic systems and devices; the range of individual responses to design problems is no less varied.

In the upper school there is less certainty about how best to enable young people to appreciate the nature of technology, its possibilities, limitations and impact on society. Some teachers feel that a conceptual structure or thematic approach is needed if the work is not to become trivialised and result in the accumulation of bits of disconnected information. One difficulty with action-based learning is that initially so much depends upon the background and early experience of the young people concerned. For this reason, one must question whether it is sufficient to plunge them raw into issues which relate to local questions of common concern, e.g. local problems of traffic flow and parking, alternative transport technologies, industrial relations and so on. Ought one first to establish some points of reference? Would the 'natural history of humanity' provide a suitable framework, as Professor Tom Stonier believes, or is society too complex to be treated in this way? It would seem worthwhile at least first to explore some aspects of man's urge to create, his use of tools and resources, his ways of organising production in home and factory and of marketing his products, his responses to and development of the environment, his changes of role and identity including family relationships and class consciousness, and his provision for education and training in agrarian, industrial and post-industrial or service-oriented societies. In the light of such considerations and through industrial archaeology or the museum service, young people might usefully explore some of the local effects of previous manifestations of man's urge to design, invent and innovate, and so be enabled to focus more sharply their own interpretations of contemporary problems and predictions of future possibilities.

The range and variety of relevant action-based learning also present a problem: to what extent can this diversity of aims and interests be seen as a recognisable pattern? Four areas can be identified. Firstly, there is the variety of experiences and exercises which can be designed to increase children's understanding of psychological, social and technological forces operating in occupational settings. Secondly, there are questions of formulating policy and policy objectives in relation to particular technological innovations, and of recognising how well these have been achieved. Thirdly, there are endless possibilities of design studies of appropriate technologies and of evaluations of alternative technologies. Fourthly, there are speculations regarding the future and the social, ecological, economic, moral and ethical criteria upon which choices are made.

The early stages of education in technological literacy in the upper school should emphasise practical understanding rather than philosophical explanation. One begins by observing technology in everyday life—in the home, school and local community—and by exploring the influence of technology on the lives of individuals, families, societies and nations, and upon those who participate in technological innovation and development or engage in industrial production. At a later stage, between the years 15 and 18, a more philosophical approach is possible; not only is it feasible to consider in practical terms how best to reorient technology and to achieve 'more with less', but it is also possible to develop skills for survival by exploring the effects of on-going technological innovations and by examining case-histories from developing countries to reveal the effect of choices upon work patterns, incomes, life-styles, patterns of training and education, and by comparing these with expectations when choices were made.

For this to happen several conditions must be fulfilled. In the first place there must be established a group (think-tank or invisible college) of creative people willing to develop the notion of technological literacy in more precise conceptual terms—possibly to suggest basic conceptual structures for the several aspects of school technology—and not only to comment on the appropriateness of these for particular stages of child development but also to suggest ways and means by which they might best be taught and learned.

Regional Centres for Science and Technology could help such a group with some aspects of the work; for example, one Centre is currently investigating the planning of an on-going and major industrial development in sociological, economic, environmental and technological terms, and examining ways and means of ensuring that the implications of this innovation are known and understood by older school children and their parents.

Secondly, it will be necessary to stake the claims of technological literacy in the context of the 'whole curriculum' and the 'common culture' arguments. Recent thinking on the curriculum, as in other fields of research, increasingly extends its objectives beyond the consideration of individual subjects to groups and larger areas of related interests. There is now a mounting concern for the curriculum as a whole and for the idea of a common culture. We would refer readers to the excellent summaries of these studies in *Schools Council Working Paper, No.* 53(1) and in Professor Lawton's *Class, Culture and the Curriculum.*(2)

Technological literacy is ideally suited to the task of giving practical and effective expression in classroom terms to these new ideas on curriculum planning. The fact that its roots are found at every level of human history and in every area of human society enables it to transcend all the natural and artificial boundaries within individuals and their social and physical environment. The strength of British education lies in the autonomy of its constituent parts and its freedom to experiment. Technology provides a means of preserving and revitalising this diversity by infusing it with unity. We thus reject the notion of the Two Cultures identified by C. P. Snow as well as the need for separate class cultures proposed by other writers.

We see unlimited scope for further development during the immediate future in an area bounded on the one hand by Professor Paul Hirst's seven discrete disciplines (mathematics, physical sciences, human sciences and history, literature and fine arts, morals, religion, and philosophy)[3]: on the other hand, by Professor Denis Lawton's five disciplines or faculties (mathematics, physical and human sciences, humanities and social science, expressive and creative arts, and morals) plus interdisciplinary work.[4] Both writers insist that all these disciplines are essential ingredients in every pupil's education and should therefore provide the framework for a common culture.

The argument about the whole curriculum has rightly centred on forms of knowledge. The Bullock Report[5] has emphasised the role of language in the acquisition, use and transmission of these forms of knowledge. In this respect, the connection between technology and language is fundamental. Through toolmaking and communal work, hand and brain formed each other and continue to form each other as language and technology through mutual interaction continue to shape and be shaped by the human species. Technology and language are the activities that call these forms of knowledge into being, bind them severally or totally together, and apply them to the venture of civilised living.

The essence of technology is in enquiry, invention and creativity but we would emphasise the message of the Bullock Report:

> Nothing has done more to confuse current educational debate than the simplistic notion that 'being told' is the polar opposite of 'finding out for oneself'. In order to accept what is offered when we are told something, we have to have somewhere to put it; and having somewhere to put it means that the framework of past knowledge and experience into which it must fit is adequate as a means of interpreting and apprehending it. Something approximating to 'finding out for ourselves' needs therefore to take place if we are to be successfully told.[6]

Technology need not exist as a separate faculty in the curriculum. It can be included in each of the disciplines and be the means of integrating them with each other and with life. In this event, each subject or discipline would be taught in such a fashion as to develop understanding of fundamental and crucial concepts by encouraging imagination, inventiveness and sheer enjoyment, with the simultaneous appreciation of the technological implications of the subject in ways suggested in preceding articles. Alternatively it can find a place in Professor

Lawton's interdisciplinary unit which could best be built in and planned as part of the whole curriculum if conceived as action-based thinking related to technological literacy. Enthusiasts for political, careers and environmental education, for leisure time and European studies, and for work experience and industrial relations might be encouraged to pool their resources in developing such a unit.

Thirdly, it would be necessary to develop a much greater degree of flexibility of organisation to operate either of these two main approaches to the promotion of technological literacy, e.g. ways of deploying staff, including team teaching, of using resources and of training teachers on site.

Time is not on our side. In 1917 A. N. Whitehead urged the need 'to eradicate the fatal disconnection of subjects which kills the vitality of our modern curriculum' and insisted that 'there is only one subject-matter for education, and that is Life in all its manifestations'.[(7)] In 1975 Professor Lawton can still say: 'One of the valid criticisms of the traditional secondary curriculum is that it has not succeeded in relating disciplines to each other nor to the real world outside schools.'[(8)]

Professor Toffler has said: 'We cannot escape the future by turning our backs on it. Foresight is uniquely human and it is essential for survival. Without this ability to imagine alternative tomorrows and to select among them, there could be no culture, no civilisation. Evolution now demands a vast enlargement of this ability, not merely among a few specialists but throughout society'.[(9)] The task is difficult and demanding. Dare we ignore it?

References

1. Schools Council Working Paper No. 53. *The Whole Curriculum* 13–16. Evans/Methuen Educational, 1975.
2. Denis Lawton, *Class Culture and the Curriculum.* Routledge & Kegan Paul, 1975.
3. The logic of the curriculum. *Journal of Curriculum Studies,* vol. 1, **2** (May 1969), p. 151. Collins.
4. *Op. cit.,* p. 88.
5. *A Language for Life.* Report of the Committee of Inquiry appointed by the Secretary of State for Education and Science under the Chairmanship of Sir Alan Bullock FBA, HMSO, 1975.
6. *Ibid.* p. 50.
7. A. N. Whitehead, *The Aims of Education,* p. 10. Benn.
8. *Op. cit.,* p. 115.
9. Alvin Toffler, *The Eco-Spasm Report,* pp. 103–4, Bantam Books, 1975.

Epilogue

MICHAEL HARRISON

Chairman SCSST, Chief Education Officer, Sheffield

and

BASIL DE FERRANTI

Vice-Chairman, SCSST

All over Europe, indeed all over the Western World, as this book is published technological change has become a central topic of discussion. For the contributors to this book this is no chance coincidence; they have been purposefully developing their thoughts for some time and may claim in their several ways to have planted and fostered the roots of the public debate now in progress. Perhaps for too long our society, in self-examination, has been absorbed in an economic mode of thought. No one of course denies the importance of attempts to define trends and to analyse causes and effects in trade and monetary systems, but the management of such systems—one hopes always for social good—is even so secondary to the mode of life of which the systems are phenomena. The numbers game of economic manipulation is in the end no substitute for action on the basic questions about quality of life in a society based on technology.

The mode of life which has concerned this book is the explosion of human talent evidenced in the modern technological society. It is a mode of life in the Western World which, whatever its faults, its consequences, its lack of control, is full of adventure, enthusiasm, artistry and in the end moral purpose—because it presses humanity to the limit of endeavour in pursuit of truth in action. And if it finds the truth, it will be because of its freedom to adventure and to go to the ultimate in any enquiry. Conversely it is unlikely that the truth will be found in any society of the totalitarian kind which obscures the scene of exploration with clouds of dogma, predetermination of political or social objective, and with the direction of the all-knowing State.

Statements about technological change tend to be dismissed as trite, even tautological. Technology implies change. Technological change has been with us, it will be said, since the first man consciously met a problem with a tool or process of his own creation. Yet it must surely now be clear that societies like ours that have come increasingly to depend upon the higher technologies for the maintenance of their large populations, for the continuous creation of their wealth and for their very morale, are approaching a qualitative change. It is no longer adequate simply to say that one more effective technology continues to succeed another—so that we build better or harness energy with less waste, for example.

Rather it is that the cumulative experience of technology may be leading us into a new world. We are perhaps moving over a watershed into a new territory liberated by a new understanding. Are we not seeing here and now the moment when enzyme technology began to be seen as an alternative to the world-wide apparatus of chemical engineering: when the iron muscle-power of the Industrial Revolution began to be superseded by the extended brain power represented by electronics and the computer: when new communications systems began to oust public transport as a means of bringing people together to get things done, so that distance disappeared as an obstacle to human progress: when the effort to rejuvenate the decaying post-industrial city began to fail for lack of possible funds and of motive and it became clear that the dispersal of urban populations into smaller centres was the happier alternative for people's quality of life: when fossil fuels ceased to be the main energy support of modern life, giving way to nuclear and solar energy? All these developments are known to moderately well-informed laymen. There will be many others equally important, some known only to experts. Together they appear to add up to a transformation of the matrix of technology and industry in which our modern civilisation is set.

It is argued in this book and elsewhere that the evidence of great change is all around us. One major new element in this evidence has, however, only just appeared, virtually overnight. It must concern every one of the present generation of young people and anyone interested in their education. This is that the context of the transition from school and further education to the world of work is radically changing. Higher technology is now insistently demanding higher qualification—and is ignoring those with lesser qualification. In parallel more young people in rapidly rising numbers at all levels of qualification, but especially in the lower ranks, are unemployed and threatened with continued unemployment. New production processes do not need the old armies of process workers. Every new design process is more streamlined than the last. But again the new processes need more people relatively in their infra-structure and more people are needed to work in service industries generally. So the flow of man power into employment at different levels of educational qualification is likely to be changing in character. And if teachers have held assumptions in the past about a neatly ordered progression from school to employment, with or without further education in between, these assumptions now need to be revised. Indeed the vulnerability of young people at the education–employment interface, the growing intractability of the problem, the dire consequences for society if the problem is not solved, mean that governments are going to have to develop—very quickly—tutelary systems to carry young people through the critical years between 16 and 25, between their leaving their basic education and their achieving firm employment in adulthood. Positive and systematic linked post-school education and training is going to be needed, allied with planned financial support for students and trainees. The relative positions therefore of government, education and training agencies and of industry and commerce will need to change, and with them the social connotations—attitudes to work (and unemployment), roles of management and unions, relationship of general to vocational education and to training, the internal structure and organisation of education and training.

The education world needs to know about the shift in patterns of life, in the life prospects for young people, and about the need for rapidly developing liaison between those responsible for education and for those overseeing the issues of manpower and employment. It has accepted for some time the warning that people must be brought up to be adaptable to the probable need to change jobs completely more than once in their working lives. Recurrent education and training systems are developing (though slowly) to help people to cope with such a mobile life style. But it is for the young person that a sudden alarm bell is ringing. To continue to educate young people as though the world were not changing fast about them could be described as social and professional blindness which future commentators, looking back at these times, would only marvel at. One must not underestimate the scale of the new task for teachers, however. A massive movement to stimulate the interpretation for them of the large changes described in this book must be launched. Some teachers, as we have read, are already pulling hard in the right direction, but the conservatism of education systems means that a widespread change in general social attitudes to the objectives and organisation of education is required if the manner of education itself is to change. It is not insignificant in this respect to note that, as these words are written, the ministers responsible for education and for employment in all the member states of the European Economic Community have instructed the appropriate directorates to prepare with the greatest urgency a report, expected to lead to a programme of recommendations to member nations, on the curricula and counselling of young people in school and their education, training and support after leaving school in the context of the newly changed environment for employment.

There is a series of clear messages for education in this book and in this epilogue. First, let us understand that the children in school now will spend the greater part of their lives in the twenty-first century not in the twentieth. Secondly, the world of the twenty-first century will be so different from the world of today that we should be unwise to make any other prediction—but we must not neglect to make it. Thirdly, today's children deserve every piece of help, every imaginative support, that we can devise to equip them for the complexities of life and society to come. Foremost among those supports is an understanding of the world of technology, swirling, evolving, burgeoning around them. The world in which they are to find their living, develop their values, express their spirits, exert their moral and political duties as citizens, serve their fellow men, is based inextricably on technology. They should be educated to understand and control for the greater good of mankind what should be a humanly and humanely orientated technology. It is the aim of this book to encourage that style of education.